# Building a Virtual Assistant for Raspberry Pi

## The practical guide for constructing a voice-controlled virtual assistant

Tanay Pant

Apress®

*Building a Virtual Assistant for Raspberry Pi*

Tanay Pant
Ghaziabad, Uttar Pradesh
India

ISBN-13 (pbk): 978-1-4842-2166-2          ISBN-13 (electronic): 978-1-4842-2167-9
DOI 10.1007/978-1-4842-2167-9

Library of Congress Control Number: 2016948437

Managing Director: Welmoed Spahr
Lead Editor: Pramila Balan
Technical Reviewer: Anand T.
Editorial Board: Steve Anglin, Pramila Balan, Laura Berendson, Aaron Black, Louise Corrigan, Jonathan Gennick, Robert Hutchinson, Celestin Suresh John, Nikhil Karkal, James Markham, Susan McDermott, Matthew Moodie, Natalie Pao, Gwenan Spearing
Coordinating Editor: Prachi Mehta
Copy Editor: Tiffany Taylor
Compositor: SPi Global
Indexer: SPi Global
Artist: SPi Global

Distributed to the book trade worldwide by Springer Science+Business Media New York, 233 Spring Street, 6th Floor, New York, NY 10013. Phone 1-800-SPRINGER, fax (201) 348-4505, e-mail orders-ny@springer-sbm.com, or visit www.springeronline.com. Apress Media, LLC is a California LLC and the sole member (owner) is Springer Science + Business Media Finance Inc (SSBM Finance Inc). SSBM Finance Inc is a **Delaware** corporation.

For information on translations, please e-mail rights@apress.com, or visit www.apress.com.

Apress and friends of ED books may be purchased in bulk for academic, corporate, or promotional use. eBook versions and licenses are also available for most titles. For more information, reference our Special Bulk Sales–eBook Licensing web page at www.apress.com/bulk-sales.

Any source code or other supplementary materials referenced by the author in this text are available to readers at www.apress.com. For detailed information about how to locate your book's source code, go to www.apress.com/source-code/. Readers can also access source code at SpringerLink in the Supplementary Material section for each chapter.

Printed on acid-free paper

*To my parents, who gave me the dream.*

# Contents at a Glance

# Contents

# About the Author

**Tanay Pant** is a writer, developer, and white hat who has a passion for web development. He contributes code to various open source projects and is the chief architect of Stock Wolf (www.stockwolf.net), a global virtual stock-trading platform that aims to impart practical education about stocks and markets. He is also an alumnus of the Mozilla Representative Program, and you can find his name listed in the credits (www.mozilla.org/credits/) of the Firefox web browser. You can also find articles written by him on web development at SitePoint and TutsPlus. Tanay acts as a security consultant and enjoys helping corporations fix vulnerabilities in their products.

# About the Technical Reviewer

**T. Anand** is a versatile technocrat who has worked on various technology projects in the last 16 years. He has also worked on industrial-grade designs; consumer appliances such as air conditioners, TVs, refrigerators, and supporting products; and uniquely developed innovative gadgets for some very specific and nifty applications, all in cross-functional domains. He offers a unique perspective with his cross-functional domain knowledge and is currently supporting product and business development and brand building for various ventures.

Anand is recognized as a Chartered Engineer by the Institute of Engineers India, Professional Engineer by the Institute of Engineers Australia, and a Lean Six-Sigma Master Black Belt by the community. He is entrepreneurial by nature and is happy to support new initiatives, ideas, ventures, startups, and nifty projects.

# Acknowledgments

I would like to express my warmest gratitude to the many people who saw me through this book and to all those who provided support, read, wrote, assisted, and offered their insights.

I would like to thank my family for their huge support and encouragement. Thank you to my father, who always inspired me to do something different, something good, with my life. I could not have asked for a better role model in my life! I am grateful to my mother, who has been the biggest source of positivity and a pillar of support throughout my life.

I want to thank Apress for enabling me to publish this book and the Apress team for providing smooth passage throughout the publishing process!

I also would like to thank the professors at the College of Technology, Pantnagar, who provided me with the support I needed to write this book. Thank you to Dr. H.L. Mandoria, Dr. Ratnesh Prasad Srivastava, Er. Sanjay Joshi, Er. Rajesh Shyam Singh, Er. B.K. Pandey, Er. Ashok Kumar, Er. Shikha Goswami, Er. Govind Verma, Er. Subodh Prasad, and Er. S.P. Dwivedi for your motivation. My deepest gratitude to all the teachers who taught me from kindergarten through engineering. Last but not the least, my thanks and appreciation go to all my friends and well wishers, without whom this book would not have been possible.

# CHAPTER 1

▉ ▉ ▉

# Introduction to Virtual Assistants

This chapter gives a detailed overview of what virtual assistants are, common virtual assistants in the market, what qualities a virtual assistant should possess, and the basic workflow and design for building a scalable virtual assistant. You also learn about the various tools required to build Melissa (your own virtual assistant) in upcoming chapters and the methodology you follow in this book.

The advent of virtual assistants has been an important event in the history of computing. Virtual assistants are useful for helping the users of a computer system automate tasks and accomplish tasks with minimum human interaction with a machine. The interaction that takes place between a user and a virtual assistant seems natural; the user communicates using their voice, and the software responds in the same way.

If you have seen the movie *Iron Man*, you can perhaps imagine having a virtual assistant like Tony Stark's Jarvis. Does that idea excite you? The movie inspired me to build my own virtual assistant software, Melissa. Such a virtual assistant can serve in the Internet of things as well as run a voice-controlled coffee machine or a voice-controlled drone.

---

**Electronic supplementary material** The online version of this chapter (doi:10.1007/978-1-4842-2167-9_1) contains supplementary material, which is available to authorized users.

© Tanay Pant 2016

T. Pant, *Building a Virtual Assistant for Raspberry Pi*, DOI 10.1007/978-1-4842-2167-9_1

# Commercial Virtual Assistants

Virtual assistants are useful for carrying out tasks such as saving notes, telling you the weather, playing music, retrieving information, and much more. Following are some virtual assistants that are already available in the market:

> *Google Now:* Developed by Google for Android and iOS mobile operating systems. It also runs on computer systems with the Google Chrome web browser. The best thing about this software is its voice-recognition ability.

> *Cortana:* Developed by Microsoft and runs on Windows for desktop and mobile, as well as in products by Microsoft such as Band and Xbox One. It also runs on both Android and iOS. Cortana doesn't entirely rely on voice commands: you can send commands by typing.

> *Siri:* Developed by Apple and runs only on iOS, watchOS, and tvOS. Siri is a very advanced personal assistant with lots of features and capabilities.

These are very sophisticated software applications that are proprietary in nature. So, you can't run them on a Raspberry Pi.

# Raspberry Pi

The software you are going to create should be able to run with limited resources. Even though you are developing Melissa for laptop/desktop systems, you will eventually run this on a Raspberry Pi.

The Raspberry Pi is a credit-card-sized, single-board computer developed by the Raspberry Pi Foundation for the purpose of promoting computer literacy among students. The Raspberry Pi has been used by enthusiasts to develop interesting projects of varying genres. In this book, you will build a voice-controlled virtual assistant named Melissa to control this little computer with your voice.

This project uses a Raspberry Pi 2 Model B. You can find information on where to purchase it at `www.raspberrypi.org/products/raspberry-pi-2-model-b/`. Do not worry if you don't currently have a Raspberry Pi; you will carry out the complete development of Melissa on a *nix-based system.

# How a Virtual Assistant Works

Let's discuss how Melissa works. Theoretically, such software primarily consists of three components: the speech-to-text (STT) engine, the logic-handling engine, and the text-to-speech (TTS) engine (see Figure 1-1).

STT                    Logic Engine                    TTS

**Figure 1-1.** *Virtual assistant workflow*

## Speech-to-Text Engine

As the name suggests, the STT engine converts the user's speech into a text string that can be processed by the logic engine. This involves recording the user's voice, capturing the words from the recording (cancelling any noise and fixing distortion in the process), and then using natural language processing (NLP) to convert the recording to a text string.

## Logic Engine

Melissa's logic engine is the software component that receives the text string from the STT engine and handles the input by processing it and passing the output to the TTS engine. The logic engine can be considered Melissa's brain; it handles user queries via a series of if-then-else clauses in the Python programming language. It decides what the output should be in response to specific inputs. You build Melissa's logic engine throughout the book, improving it and adding new functionalities and features as you go.

## Text-to-Speech Engine

This component receives the output from Melissa's logic engine and converts the string to speech to complete the interaction with the user. TTS is crucial for making Melissa more humane, compared to giving confirmation via text.

This three-component system removes any physical interaction between the user and the machine; the users can interact with their system the same way they interact with other human beings. You learn more about the STT and TTS engines and how to implement them in Chapter 2.

From a high-level view, these are the three basic components that make up Melissa. This book shows you how to do all the necessary programming to develop them and put them together.

# Setting Up Your Development Environment

This is a crucial section that is the foundation of the book's later chapters. You need a computer running a *nix-based operating system such as Linux or OS X. I am using a MacBook Air (early 2015) running OS X 10.11.1 for the purpose of illustration.

## Python 2.x

You will write Melissa's code in the Python programming language. So, you need to have the Python interpreter installed to run the Python code files. *nix systems generally have Python preinstalled. You can check whether you have Python installed by running the following command in the terminal of your operating system:

```
$ python --version
```

This command returns the version of the Python installed on your system. In my case, it gives the following output:

```
Python 2.7.11
```

This should also work on other versions of Python 2.

---

■ **Note**  I am using Python 2 instead of Python 3 because the various dependencies used throughout the book are written in Python 2.

---

## Python Package Index (PyPI)

You need pip to install the third-party modules that are required for various software operations. You use these third-party modules so you do not have to reinvent the wheels of assorted basic software processes.

You can check whether pip is installed on your system by issuing the following command:

```
$ pip --version
```

In my case, it gives this output:

```
pip 7.1.2 from /usr/local/lib/python2.7/site-packages (python 2.7)
```

If you do not have pip installed, you can install it by following the guide at https://pip.pypa.io/en/stable/installing/.

## Version Control System (Git)

You use Git for version control of your software as you work on it, to avoid losing work due to hardware failure or system administrator mistakes. You can use GitHub to upload your Git repository to an online server. You can check whether you have Git installed on your system by issuing the following command:

```
$ git --version
```

This command gives me the following output:

```
git version 2.6.2
```

If you do not have Git installed, you can install it using the instructions at http://git-scm.com/downloads.

## PortAudio

PortAudio is an open source input/output library. It is cross platform and is available in the form of source files that can be downloaded from www.portaudio.com/download.html. It can be compiled on many platforms such as Windows, OS X, and Unix. PortAudio provides a simple API for recording and playing sound that is used by some of the speech-recognition modules in future chapters.

# PyAudio

PyAudio provides Python bindings for PortAudio. With the help of this software, you can easily use Python to record and play audio on a variety of platforms, which is exactly what you need for your STT engine. You can find the instructions for installing PyAudio at http://people.csail.mit.edu/hubert/pyaudio/.

You also need a microphone via which you can speak to your computer (and perform voice recording) and speakers to hear the output. Most modern laptops have these installed by default. For a Raspberry Pi, you need an external microphone and speakers/earphones.

# Designing Melissa

You will follow the DRY (don't repeat yourself) and KISS (keep it simple, stupid) principles and use modular code to design Melissa. Doing so helps maintain your code properly and makes it easier to scale the code in the future when you want to add cool features to your existing codebase. So, let's first design the structure of your code directories:

```
gitignore
GreyMatter/
    SenseCells/
        __init__.py
```

```
    ...
  __init__.py
    ...
main.py
brain.py
profile.yaml.default
requirements.txt
```

In this directory structure, ... denotes that files will be added here in the future as you go through the chapters in this book. The folders containing __init__.py files are Python packages. The main.py file will be entry point of the program and will contain the source code for the completed STT engine; it will pass commands (in the form of strings) to brain.py for handling (this is the logic engine I previously mentioned). The SenseCells package will contain the TTS engine, and the GreyMatter package will contain the various mini-features that can be integrated into the software as you progress through the book. requirements.txt file will be used for keeping tabs on the third-party Python modules you use in this project.

The profile.yaml.default file will store information such as the name of the user as well as the city where the user lives, in YAML format. The profile.yaml file is crucial for executing the main.py file. The user will issue the following to get this software up and running:

```
$ cp profile.yaml.default profile.yaml
```

You append the .default suffix so that if users put personal information in the profile.yaml file and create a pull request on GitHub, it won't include their private changes to the profile.yaml file, because it is mentioned in the .gitignore file.

Currently the contents of profile.yaml.default are as follows:

```
name:
  Tanay
city_name:
  New Delhi
```

The contents of the .gitignore file are as follows:

```
profile.yaml
*.pyc
```

Now that you know the high-level directory structure of the project, you can go ahead and create the skeleton structure. This structure will help you keep the code base clean and properly organized as you move through the book and work on building new features.

# Learning Methodology

This section describes the methodology you use throughout the book: understanding concepts, learning by prototyping, and then developing production-quality code to integrate into the skeleton structure you just developed (see Figure 1-2).

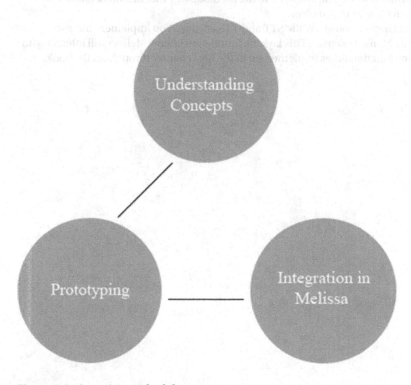

***Figure 1-2.*** *Learning methodology*

First you explore the theoretical concepts as well as understand the core principles that will enhance your creativity and help you see different ways to implement features. This part may seem boring to some people, but do not skip these bits.

Next, you implement your acquired knowledge in Python code and play around with it to convert your knowledge into skills. Prototyping will help you to understand the functioning of individual components without the danger of messing up the main codebase. Finally, you edit and refactor the code to create good-quality code that can be integrated with the main codebase to enhance Melissa's capabilities.

# Summary

In this chapter, you learned about what virtual assistants are. You also saw various virtual assistants that exist in the commercial market, the features a virtual assistant should possess, and the workflow of a voice-controlled virtual assistant. You designed Melissa's codebase structure and were introduced to the methodology that this book follows to create an effective learning workflow.

In the next chapter, you study the STT and TTS engines. You implement them in Python to create Melissa's senses. This lays the foundation of how Melissa will interact with you; you use the functionalities implemented in the next chapter throughout the book.

**CHAPTER 2**

■ ■ ■

# Understanding and Building an Application with STT and TTS

This chapter introduces you to the concepts of speech-to-text (STT) and text-to-speech (TTS). It discusses various STT engines, and you build a Python program that records audio. You then graduate to an application that converts whatever you say to text. You also look at the use of various TTS engines and implement them to make a program that repeats whatever you say.

## Speech-to-Text Engines

As you saw in Chapter 1, the STT engine is one of the three main components of the virtual assistant, Melissa. This component is the entry point for the software's control flow. Hence, you need to incorporate this piece of code into the main.py file. First, you need a sophisticated STT engine to use for Melissa. Let's look at the various STTs available on the Web for free use with your application

### Freely Available STTs

Some of the best STTs available on the Internet are as follows:

- *Google STT* is the STT system developed by Google. You may already have used the Google STT if you have an Android smartphone, because it is used in Google Now. It has one of the best recognition rates. But it can only transcribe a limited amount of speech per day (API limitation) and needs an active Internet connection to work.

© Tanay Pant 2016

T. Pant, *Building a Virtual Assistant for Raspberry Pi*, DOI 10.1007/978-1-4842-2167-9_2

- *Pocketsphinx* is an open source speech decoder developed under the CMU Sphinx Project. It is quite fast and has been designed to work well on mobile operating systems such as Android as well as embedded systems (like Raspberry Pi). The advantage of using Pocketsphinx is that the speech recognition is performed offline, which means you don't need an active Internet connection. However, the recognition rate is nowhere close to that of Google's STT.

- *AT&T STT* was developed by AT&T. The recognition rate is good, but it needs an active connection to work, just like Google STT.

- *Julius* is a high-performance, open source speech-recognition engine. It does not need an active Internet connection, like Pocketsphinx. It is quite complicated to use because it requires the user to train their own acoustic models.

- *Wit.ai STT* is a cloud-based service provided to users. Like AT&T and Google STT, it requires an active Internet connection to work.

- *IBM STT* was developed by IBM and is a part of the Watson division. It requires an active Internet connection to work.

This project uses Google STT because it is one of the most accurate STT engines available. In order to use Google STT in your project, you need a Python module called `SpeechRecognition`.

## Installing SpeechRecognition

You install `SpeechRecognition` by issuing the following command via the terminal:

```
$ pip install SpeechRecognition
```

This sets up the `SpeechRecognition` module for you. This library supports Google Speech Recognition, Wit.ai, IBM Speech to Text, and AT&T Speech to Text. You can choose any of these for your version of Melissa.

# Recording Audio to a WAV File

Let's write a small Python program to see how this library works. This program records the user's voice and saves it to a `.wav` file. Recording the audio to a WAV file will help you get comfortable with the `SpeechRecognition` library. You also use this method of recording speech to a WAV file and then passing that file to the STT server in Chapter 8:

```
import speech_recognition as sr

r = sr.Recognizer()
with sr.Microphone() as source:
```

```
print("Say something!")
audio = r.listen(source)

with open("recording.wav", "wb") as f:
    f.write(audio.get_wav_data())
```

Let's examine this program line by line. The first statement imports the SpeechRecognition module as sr. The second block of code obtains the audio from the microphone. For this purpose, it uses the Recognizer() and Microphone() functions. This example uses PyAudio because it uses the Microphone class. The third block of code writes the audio to a WAV file named recording.wav.

Run this file from the terminal. You should get the results you expect: whatever you said into the microphone was recorded to recording.wav. Notice that the Python program stops recording when it detects a pause in your speech for a certain amount of time.

Running the program on my system gave me the output shown in Figure 2-1 and in the following snippet. Your Python program produces the recording.wav file. You may also receive a warning message like the one you can see on my console—if so, don't worry about it, because it does not effect the working of your program. Here's my output:

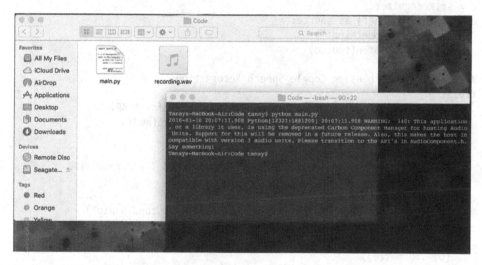

***Figure 2-1.*** *Recording to a WAV file: console output*

```
Tanays-MacBook-Air:Melissa-Core-master tanay$ python main.py
2016-01-10 20:07:11.908 Python[12321:1881200] 20:07:11.908 WARNING:  140:
This application, or a library it uses, is using the deprecated Carbon
Component Manager for hosting Audio Units. Support for this will be removed
in a future release. Also, this makes the host incompatible with version 3
audio units. Please transition to the API's in AudioComponent.h.
Say something!
```

Great! Now you understand the basics of working with the SpeechRecognition library. If for some reason the speech recording is not working for you, you may want to skip to Chapter 8 to follow a web-based approach for capturing the user's voice, and then continue from this chapter.

# Speech Recognition

Let's now get to the code that records the audio and sends it to the STT for conversion to a text string. The page of the SpeechRecognition module at PyPi has a link to a code sample that performs the STT conversion. This section discusses that example.

## Google STT

Take a look at this new code snippet:

```
import speech_recognition as sr
# obtain audio from the microphone
r = sr.Recognizer()
with sr.Microphone() as source:
    print("Say something!")
    audio = r.listen(source)

# recognize speech using Google Speech Recognition
try:
    # for testing purposes, you're just using the default API key
    # to use another API key, use `r.recognize_google(audio,
    key="GOOGLE_SPEECH_RECOGNITION_API_KEY")`
    # instead of `r.recognize_google(audio)`
    print("Google Speech Recognition thinks you said " + r.recognize_
    google(audio))
except sr.UnknownValueError:
    print("Google Speech Recognition could not understand audio")
except sr.RequestError as e:
    print("Could not request results from Google Speech Recognition service;
    {0}".format(e))
```

First you use the microphone as the source to listen to the audio and use the same code snippet that you used when you recorded the audio file. This snippet uses a try/except clause for error handling. If the error is sr.UnknownValueError, the program returns "Google Speech Recognition could not understand audio". If you get a sr.RequestError error, you take its value in e and print "Could not request results from Google Speech Recognition service" along with the technical details of the error returned by Google STT. In the try clause, you use the r.recognize_google() function to pass the audio as an argument to Google STT. It then prints out what you said, as interpreted by Google, in the form of a string. This method uses the default API key; you do not need to enter a unique key for development purposes.

> ■ **Note** You can find instructions for how to obtain the Speech API keys from Google on the Chromium web site: https://www.chromium.org/developers/how-tos/api-keys.

## Wit.ai STT

If you wish to use Wit.ai STT, use this snippet in place of the try/except clause used in the previous code:

```
# recognize speech using Wit.ai
WIT_AI_KEY = "INSERT WIT.AI API KEY HERE"

try:
    print("Wit.ai thinks you said " + r.recognize_wit(audio, key=WIT_AI_KEY))
except sr.UnknownValueError:
    print("Wit.ai could not understand audio")
except sr.RequestError as e:
    print("Could not request results from Wit.ai service; {0}".format(e))
```

While using the Wit.ai service, you have to obtain the Wit.ai key stored in the WIT_AI_KEY constant. You use the r.recognize_wit() function to pass the audio and the key as arguments.

## IBM STT

To use IBM STT, use the following code snippet:

```
# recognize speech using IBM Speech to Text
IBM_USERNAME = "INSERT IBM SPEECH TO TEXT USERNAME HERE"
IBM_PASSWORD = "INSERT IBM SPEECH TO TEXT PASSWORD HERE"

try:
    print("IBM Speech to Text thinks you said " + r.recognize_ibm(audio,
    username=IBM_USERNAME, password=IBM_PASSWORD))
except sr.UnknownValueError:
    print("IBM Speech to Text could not understand audio")
except sr.RequestError as e:
    print("Could not request results from IBM Speech to Text service;
    {0}".format(e))
```

When using the IBM STT service, you have to obtain an IBM STT username and password, which you assign to the IBM_USERNAME and IBM_PASSWORD constants, respectively. You then invoke the r.recognize_ibm() function and pass the audio, username, and password as arguments.

## AT&T STT

To use AT&T STT, use the following code snippet:

```
# recognize speech using AT&T Speech to Text
ATT_APP_KEY = "INSERT AT&T SPEECH TO TEXT APP KEY HERE"
ATT_APP_SECRET = "INSERT AT&T SPEECH TO TEXT APP SECRET HERE"

try:
    print("AT&T Speech to Text thinks you said " + r.recognize_att(audio,
    app_key=ATT_APP_KEY, app_secret=ATT_APP_SECRET))
except sr.UnknownValueError:
    print("AT&T Speech to Text could not understand audio")
except sr.RequestError as e:
    print("Could not request results from AT&T Speech to Text service;
    {0}".format(e))
```

To use the AT&T STT service, you have to obtain an AT&T app key as well as an app secret and assign them to the ATT_APP_KEY and the ATT_APP_SECRET constants, respectively. You then have to implement the r.recognize_att() function and pass audio, app_key, and app_secret as arguments.

# Melissa's Inception

As you may have noticed, the SpeechRecognition package provides a very nice, generic wrapper that lets developers incorporate a wide variety of online STTs into applications. Go ahead and run the speech-recognition program.

As expected, the following snippet shows that the program took what I said into the microphone, recognized it, converted it into a string, and displayed it on the terminal. In this case, I said, "hi Melissa how are you":

```
Tanays-MacBook-Air:Melissa-Core-master tanay$ python main.py
2016-01-10 20:49:11.192 Python[12460:1899626] 20:49:11.191 WARNING:  140:
This application, or a library it uses, is using the deprecated Carbon
Component Manager for hosting Audio Units. Support for this will be removed
in a future release. Also, this makes the host incompatible with version 3
audio units. Please transition to the API's in AudioComponent.h.
Say something!
Google Speech Recognition thinks you said hi Melissa how are you
```

Wonderful! You have now programmed the first of the three components required to build a functional virtual assistant. You can speak to your computer, and you can be rest assured that whatever you say will be converted to a string.

# Text-to-Speech Engine

Let's turn now to the third component of the virtual assistant abstract system: text-to-speech. A virtual assistant does not feel human if it replies to queries in the form of text output like that in the terminal application. You need Melissa to talk; and for that purpose, you need to use a TTS engine.

Different types of TTS are available for different platforms. Because TTS is native software that is OS dependent, this section discusses the software available for OS X and Linux-based systems, both of which are *nix-based. It is perfectly possible to program on a Raspberry Pi from the beginning, but for the sake of learning and testing, I am working on the laptop, as you may be, too. This approach allows you to work your way through the book even if you don't have a Raspberry Pi or if the Raspberry Pi you have ordered hasn't arrived just yet.

## OS X

OS X comes preloaded with the say command, which allows you to access the built-in TTS without having to install any additional third-party software. The voice quality and dialect of say are among the best, and the response seems quite human and realistic.

To test the say command, open the command line and enter the following command:

```
$ say "Hi, I am Melissa"
```

If you have your speakers turned on or if you are listening via earphones, you can listen to your system speak these words out loud to you.

## Linux

Some Linux distributions come with software called eSpeak preinstalled. However, other distributions, like Linux Mint, do not have eSpeak preinstalled. You can find the instructions to install the eSpeak utility on your system at http://espeak.sourceforge.net.

Once you have installed the eSpeak software, you can test it via the terminal by entering the following command:

```
$ espeak "Hi, I am Melissa"
```

This causes your system to speak whatever you have written. Note that eSpeak is not as impressive as OS X's say command; the voice quality is robotic and has a strange accent. Despite this, I have included eSpeak because of its small size. You can use any other TTS engine if you want to and edit the code of the TTS engine that you write shortly accordingly.

# Building the TTS Engine

To make your software cross-platform between OS X and Linux, you have to determine which OS your software is running on. You can find that out by using sys.platform in Python. The value of sys.platform on Apple systems is Darwin, and on Linux-based systems it is either linux or linux2.

Let's write the Python code to accomplish the task:

```python
import os
import sys

def tts(message):
    """
    This function takes a message as an argument and converts it to speech
    depending on the OS.
    """
    if sys.platform == 'darwin':
        tts_engine = 'say'
        return os.system(tts_engine + ' ' + message)
    elif sys.platform == 'linux2' or sys.platform == 'linux':
        tts_engine = 'espeak'
        return os.system(tts_engine + ' "' + message + '"')
```

Let's go through the code. The first two import statements import the os and sys modules. Then you define a function called tts that takes a message as an argument. The if statement determines whether the platform is OS X; then it assigns the say value to the tts_engine variable and returns os.system(tts_engine + ' ' + message). This executes the say command with the message on the terminal. Similarly, if the platform is Linux based, it assigns espeak to the tts_engine variable.

To test the program, you can add the following additional line at the bottom of the code:

```python
tts("Hi handsome, this is Melissa")
```

Save the code, and run the Python file. It should execute successfully.

## Repeat What I Say

For the sake of exercise and fun, construct a Python program that detects whatever you say and repeats it. This involves a combination of the STT and TTS engines. You have to make the following assignment:

```python
message = r.recognize_google(audio)
```

# Integrating STT and TTS in Melissa

As discussed in Chapter 1, you are past the stages of learning concepts and prototyping STT and TTS; now it's time to integrate the STT engine as well as the TTS engine in Melissa in a proper, reusable fashion.

First, let's put the TTS in place, because the TTS engine is complete and does not require any changes or additions to the code. Put this in a file called tts.py, and place it in the following location:

```
GreyMatter/
       SenseCells/
              __init__.py
              tts.py
       __init__.py
       ...
```

You may remember this directory structure from the "Designing Melissa" section of Chapter 1. Now the TTS is in a package and can be called from other Python files after importing it. Next, edit the main.py Python file:

```python
import sys

import yaml
import speech_recognition as sr

from GreyMatter.SenseCells.tts import tts

profile = open('profile.yaml')
profile_data = yaml.safe_load(profile)
profile.close()

# Functioning Variables
name = profile_data['name']
city_name = profile_data['city_name']

tts('Welcome ' + name + ', systems are now ready to run. How can I help you?')

def main():
    r = sr.Recognizer()
    with sr.Microphone() as source:
        print("Say something!")
        audio = r.listen(source)

    try:
        speech_text = r.recognize_google(audio).lower().replace("'", "")
        print("Melissa thinks you said '" + speech_text + "'")
    except sr.UnknownValueError:
        print("Melissa could not understand audio")
```

```
    except sr.RequestError as e:
        print("Could not request results from Google Speech Recognition
        service; {0}".format(e))

    tts(speech_text)

main()
```

This program performs the same function as the "Repeat What I Say" program you created earlier, but it has a much more modular, cleaner approach. You can easily add more features when you have new ideas, without having to touch existing code.

Let's study the changes that have been made in the main.py file as compared to what you had earlier. Notice that a new package named yaml has been imported. You have also imported the tts function so that it can be used in the main file.

This is used to parse the profile.yaml file you created in Chapter 1. You open the YAML file and use the yaml.safe_load() function to load data from the file and save it to profile_data. You then close the file you opened. You can retrieve the data in the form of profile_data['name'] and assign it to appropriate variables for use in the future.

You then call the tts function imported from GreyMatter.SenseCells.tts to include a welcome note for the user. If the user has customized the configuration in the profile.yaml file, it uses their name in the welcome note. The entire STT is placed in a function called main, and that function is called at the end of the code. This completes your construction of two out of three components of the virtual assistant.

# Version-Controlling Your Source Code

Because you have finished building all the necessary components for this chapter, let's version-control your source code. Start by initializing an empty Git repository by entering the following command:

```
$ git init
```

Now, check the status of the added/modified files, add the files, and commit them:

```
$ git status
$ git add --all
$ git commit -m "Add STT and TTS functionality"
```

You can view all the previous commit messages by entering the following command in the terminal:

```
$ git log --pretty=oneline
```

You have successfully committed the first version of changes into your local Git repository. You can also push your changes if you have a repository for this purpose on GitHub. If not, you can create an empty repository at GitHub, and it will give you the directions to upload your local Git repository.

# Obtaining the Code from GitHub

I have uploaded a completed version (completed in the sense of the chapters of this book) of Melissa at GitHub. You can access the code from `https://github.com/Melissa-AI/Melissa-Core` (see Figure 2-2).

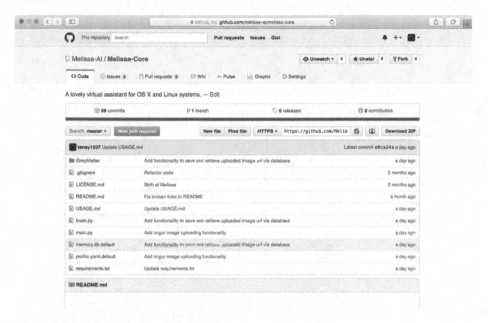

*Figure 2-2.* *Melissa's codebase at GitHub*

You can fork this repository by clicking the Fork button at upper right. Then you can clone your fork to get Melissa running locally.

You can create pull requests whenever you wish to make changes to Melissa's official repository to either fix bugs or add features. Make sure you first create an issue before fixing any bug that requires extensive code changes and before working to develop a new feature, because this will let others know that you are working on it and there won't be duplicates.

# Summary

In this chapter, you learned about some of the widely used STT and TTS engines, and you used the freely available STT and TTS engines to create a program in Python that can record what the user is saying and repeat it. Then you integrated this code into Melissa so that she can listen as well as talk. Finally, you version-controlled your source code so that you can share your code on GitHub.

In the next chapter, you learn about building the third component of a virtual assistant: the logic engine to make Melissa smarter. You build a conversation module so you can converse with Melissa.

# CHAPTER 3

■ ■ ■

# Getting Your Hands Dirty: Conversation Module

In this chapter, you learn how to implement a conversation module to make Melissa understand what you are saying, with the help of a Python program that implements keyword-recognition techniques. You refine the code of the program to make it more efficient, so that you can have a general conversation with Melissa and ask questions like, "How are you?" and "Who are you?"

You have reached the step of building a virtual assistant that involves designing a logic engine. Melissa is basically a parrot right now, repeating what you say. This assistant needs to be more than that; it needs to understand what you say. In a quest to make Melissa smart, let's design a conversation module.

Before you learn how to implement this module in Python, let's revisit the code skeleton from Chapter 1 and see how you build and add components of the logic engine, keeping the different modules isolated from each other. You have already incorporated the STT and TTS in the code skeleton, so in this chapter you immediately implement the code you develop into the project instead of prototyping.

## Logic Engine Design

main.py is the STT engine of your software, and it is also the entry point to your program. You need main.py to direct user queries to its logic engine, which you code in the brain.py file. The brain.py file will contain a ladder of if/else clauses to determine what the user wants to say. If there is a pattern match with one of the statements, brain.py call the corresponding module.

Figure 3-1 shows the control flow of the program. This will be similar for all the modules you develop for Melissa in future chapters. The difference will be that some other module is called by brain.py instead of general_conversations.py.

The GreyMatter package will hold logic-engine modules that you build to make Melissa smarter in the future, such as a weather module, opening a web site, playing music, and so on. The GreyMatter package also contains the general_conversations.py file.

© Tanay Pant 2016
T. Pant, *Building a Virtual Assistant for Raspberry Pi*, DOI 10.1007/978-1-4842-2167-9_3

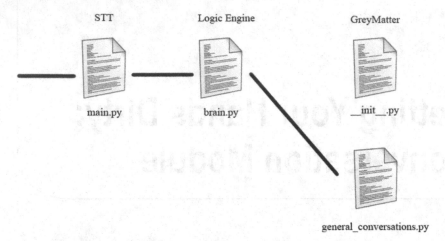

*Figure 3-1.* *Logic engine design*

# Making Melissa Responsive

Let's get to the task of making Melissa responsive, so that she can respond to questions. This requires you to compare the speech_text variable to a predefined string.

First, create the general_conversations.py file in the GreyMatter folder, and program it as follows:

```
from SenseCells.tts import tts

def who_are_you():
    message = 'I am Melissa, your lovely personal assistant.'
    tts(message)

def undefined():
    tts('I dont know what that means!')
```

Let's go through the code. In the first statement, you import the tts function from the SenseCells.tts package. You then write an elementary function, who_are_you(), in which a reply string is assigned to the variable message. This message is then spoken by the tts function. The undefined() function is called whenever the brain cannot find a match; it's called from the final else statement.

For now, let's keep general_conversations.py short for the sake of illustration. Later, you revisit this file to add features to it and improve the code.

It's time to design the brain function in the brain.py file:

```
from GreyMatter import general_conversations

def brain(name, speech_text):
    def check_message(check):
```

```
    if speech_text ==   check:
        return True
    else:
        return False

if check_message('who are you'):
    general_conversations.who_are_you()
else:
    general_conversations.undefined()
```

In the first statement, you import general_conversations from the GreyMatter package. You then define a function called brain that takes name and speech_text as arguments (you use the name argument later in this chapter). Inside it is another function named check_message() that takes check as an argument. This function compares two strings, speech_text and check, to see if they are equal. Then the function returns either True (if the string matches) or False (if it doesn't).

Going further down the code, you find the if/else ladder. You invoke the check_message() function with 'who are you' as the argument to see if this is what the user said. If True, you call the who_are_you() function from general_conversations. If False, then you fall back to the undefined() function. You revisit this file later to edit the code and improve check_message().

Finally, you need to make changes to main.py so that you can pass the user's speech to the brain function:

```
import sys

import yaml
import speech_recognition as sr

from brain import brain
from GreyMatter.SenseCells.tts import tts

profile = open('profile.yaml')
profile_data = yaml.safe_load(profile)
profile.close()

# Functioning Variables
name = profile_data['name']
city_name = profile_data['city_name']

tts('Welcome ' + name + ', systems are now ready to run. How can I help you?')

def main():
    r = sr.Recognizer()
    with sr.Microphone() as source:
        print("Say something!")
        audio = r.listen(source)
```

```
    try:
        speech_text = r.recognize_google(audio).lower().replace("'", "")
        print("Melissa thinks you said '" + speech_text + "'")
    except sr.UnknownValueError:
        print("Melissa could not understand audio")
    except sr.RequestError as e:
        print("Could not request results from Google Speech Recognition
        service; {0}".format(e))

    brain(name, speech_text)
main()
```

First, you import the `brain()` function from `brain.py`. Inside the `main()` function, you add `brain()`, where you pass the `name` and `speech_text` arguments.

Your program is now ready to run and to be tested. Go to the terminal, and start the program by issuing the following command:

```
$ python main.py
```

When you see the "Say something!" message, say "Who are you?" into the microphone. You should get the following reply: "I am Melissa, your lovely personal assistant." Try saying something else, and you should receive the following reply: "I dont know what that means!"

There are two problems with the existing system:

1.  The library of conversation clauses is limited and static.

2.  The recognition system in the brain is very poor, because it compares strings.

You can solve the first problem by having an array of messages and using the `random.choice()` function to answer the user's question. The second problem is much more complex in nature. Even if the user says something like, "Hey, who are you?" the logical engine will pass control to the `undefined()` function. This shouldn't be the case, because "Who are you?" and "Hey, who are you?" essentially mean the same thing. This problem can be handled by checking `speech_text` for certain keywords.

# Fixing Limitation 1

Let's edit `general_conversations.py` to implement the fix just discussed and include some new conversation snippets:

```
import random
from SenseCells.tts import tts

def who_are_you():
    messages = ['I am Melissa, your lovely personal assistant.',
    'Melissa, didnt I tell you before?',
```

```
    'You ask that so many times! I am Melissa.']
    tts(random.choice(messages))

def how_am_i():
    replies =['You are goddamn handsome!', 'My knees go weak when I see you.',
    'You are sexy!', 'You look like the kindest person that I have met.']
    tts(random.choice(replies))

def tell_joke():
    jokes = ['What happens to a frogs car when it breaks down? It gets toad
    away.', 'Why was six scared of seven? Because seven ate nine.', 'No, I
    always forget the punch line.']
    tts(random.choice(jokes))

def who_am_i(name):
    tts('You are ' + name + ', a brilliant person. I love you!')

def where_born():
    tts('I was created by a magician named Tanay, in India, the magical land
    of Himalayas.')

def how_are_you():
    tts('I am fine, thank you.')

def undefined():
    tts('I dont know what that means!')
```

To take care of the static replies, you import the random module on the
first line. You then make an array of appropriate replies to a particular question and pass
random.choice(array_of_appropriate_messages) to the tts function. This causes the
virtual assistant to give different answers to a question each time the question is asked.
You also add some other questions that people may feel inclined to ask a virtual assistant.
You can find the code for general_conversations.py on GitHub: https://github.com/
Melissa-AI/Melissa-Core/blob/master/GreyMatter/general_conversations.py.

# Fixing Limitation 2

To fix the second limitation discussed earlier, edit the code in the brain.py file:

```
from GreyMatter import general_conversations

def brain(name, speech_text):
    def check_message(check):
        """
        This function checks if the items in the list (specified in
        argument) are present in the user's input speech.
        """
```

```
        words_of_message = speech_text.split()
        if set(check).issubset(set(words_of_message)):
            return True
        else:
            return False

if check_message(['who','are', 'you']):
    general_conversations.who_are_you()

elif check_message(['how', 'i', 'look']) or check_message(['how', 'am', 'i']):
    general_conversations.how_am_i()

elif check_message(['tell', 'joke']):
    general_conversations.tell_joke()

elif check_message(['who', 'am', 'i']):
    general_conversations.who_am_i(name)

elif check_message(['where', 'born']):
    general_conversations.where_born()

elif check_message(['how', 'are', 'you']):
    general_conversations.how_are_you()

else:
    general_conversations.undefined()
```

The main change in this file is the code edit in the check_message() function
(in addition to the additions of the conversation snippets in the if/else ladder). Let's
analyze the changes in check_message. First, you split the speech_text string and store it
in a variable called words_of_message. This results in an array of words that are present in
the speech.

Note that the check argument in the updated brain.py file refers to an array of
strings (not a string, as in the previous version). You then make a set of check and
words_of_message, which removes any duplicate words. Finally, you check whether the
set check is a subset of the set words_of_message. If it is a subset, then it returns True;
otherwise, it returns False (see Figure 3-2).

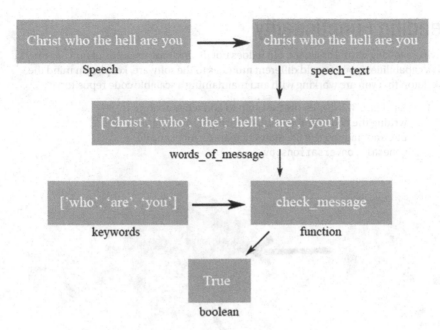

*Figure 3-2. Keyword-detection scheme*

Now you have to learn how to use this function to improve your recognition rates. Let's consider the example of the question "Who are you?" Here are some possible variations of the question:

- "Hey, who are you?"

- "Can you tell me who are you?"

- "I wish to know who you are!"

- "For Christ's sake tell me who the hell you are!"

This list is by no means comprehensive, but it is representative of the various ways a user can ask Melissa the question. Notice that three words are present in all the statements: *who, are,* and *you*. Hence, it can be considered a safe bet to make an array of these three keywords and consider it an identifier for the base question, "Who are you?"

The construction for checking the speech for this base question is as follows:

```
check_message(['who','are', 'you']):
```

Similarly, you can extend Melissa's answering capability by adding a function to `general_conversations.py` and including the corresponding check in `brain.py`.

Congratulations—you have successfully built your talking virtual assistant!

# Extending Functionality

You having a talking virtual assistant, but it does not do anything useful as of now. To extend Melissa's capabilities, you can add different modules to the software, keeping in mind the code skeleton that you are working with and maintaining a scalable code repository:

1. As Figure 3-3 shows, the first step to add a new feature is writing the code for the feature, storing it in a separate file, and naming the file after the feature: for example, general_conversations.py.

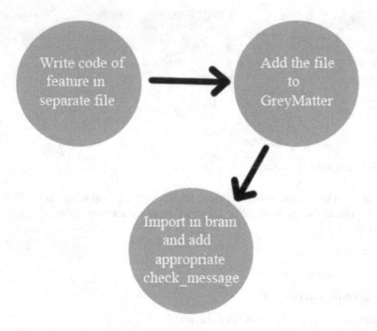

*Figure 3-3.* *Extending functionality*

2. Add the file to the GreyMatter package, from which it can be imported as a module to be used in brain.py.

3. Import the module in brain.py, and add appropriate checks for keywords related to your functionality with the help of the check_message() function. If check_message() returns true, execute the feature.

You follow the same methodology to add new features and functionality in future chapters. Let's look at an example of how you can follow this process to write a new basic feature for Melissa.

# What's the Time, Melissa?

Let's write a new module for telling the time. It will allow you to ask Melissa the time whenever you want. Create a new file in the GreyMatter folder, named tell_time.py, and add the following code to it:

```
from datetime import datetime

from SenseCells.tts import tts

def what_is_time():
    tts("The time is " + datetime.strftime(datetime.now(), '%H:%M:%S'))
```

In this code, you first import the datetime module and the datetime function and call datetime.now() in the tts function, in the what_is_time() function. Also note that you format the time using datetime.strftime(), using the format "x hours, y minutes, z seconds."

Next, make the following changes in brain.py to implement this feature:

```
from GreyMatter import tell_time, general_conversations
```

This change needs to be made in the first line to import the tell_time module. Now add this elif clause under the brain() function:

```
elif check_message(['time']):
        tell_time.what_is_time()
```

Here, the keyword is 'time', and recognizing this keyword via the check_message() function causes the execution of the feature. Go ahead and execute this program to ask Melissa the time.

# Committing Changes

It is time to commit the changes you have made to Melissa to the Git repository. Enter the following commands in your terminal:

```
$ git status
$ git diff
$ git add --all
$ git commit -m "Add conversation and time modules"
$ git push
```

The git diff command shows you the changes in the individual files that have taken place since the last commit. This will help you to review the changes.

# Summary

In this chapter, you learned to make the logical engine of a virtual assistant and develop a text-recognition system so that the assistant can understand the meaning of your commands. You then developed a conversation module that helps you hold conversations with Melissa and a time module that enables Melissa to tell the time when asked.

In the next chapter, you learn to gather data from the Internet to make Melissa smarter and more useful. You see how to scrape business news from the Internet so that Melissa can read news to you, tell you the weather, and define things from Wikipedia.

**CHAPTER 4**

■ ■ ■

# Using the Internet to Gather Information

In this chapter, you learn about using information from the Internet to make Melissa more interactive. You create modules for getting weather information, defining keywords from Wikipedia, and mining data from various web sites (such as news web sites). You then implement these modules in Python to make Melissa more useful.

The Internet is a mine full of loads of useful information and, well, useless information. Some of the meaningful, useful information you can retrieve from the Internet includes weather, definitions from Wikipedia, and business news. It would be really useful to have a virtual assistant that can provide you with all this information on demand. Let's construct these functionalities in Melissa's GreyMatter.

The three information-retrieval features—telling you the weather, reading definitions from Wikipedia, and retrieving business news—are of increasing code complexity, from very easy to intermediate. For the first two features, you use third-party modules that you install using pip. For the third example, you mine data and parse it to get meaningful information from a news web site.

## How's the Weather?

In this section, you obtain weather information from the weather.com web site with the help of a Python module name pywapi. To install pywapi, enter the following command in your terminal:

```
$ pip install pywapi --allow-external --allow-unverified
```

pywapi requires the city code of your city in order to retrieve the weather report. To find the city code, open the weather.com web site, search for your city name, and look at the URL. I searched for my city name, New Delhi, and got the following URL:

```
http://www.weather.com/weather/today/l/INXX0096:1:IN
```

© Tanay Pant 2016
T. Pant, *Building a Virtual Assistant for Raspberry Pi*, DOI 10.1007/978-1-4842-2167-9_4

Figure 4-1 shows the part of the URL to note for future use.

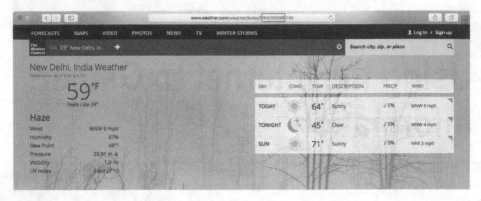

***Figure 4-1.*** *Getting a city code at weather.com*

Here, INXX0096 is the city code required by the pywapi module. Let's log this information in the profile.yaml file:

```
city_code:
    INXX0096
```

Now it is time to import this piece of information into the main.py file, which you do by adding the following line of code:

```
city_code = profile_data['city_code']
```

You also need to pass this information to the logical engine (a.k.a. the brain), so the brain() function in the main() function must be modified as well in main.py:

```
brain(name, speech_text, city_name, city_code)
```

Before diving into brain.py to make the edits there, first create a file named weather.py in the GreyMatter folder and add the following code to it:

```
import pywapi

from SenseCells.tts import tts

def weather(city_name, city_code):
    weather_com_result = pywapi.get_weather_from_weather_com(city_code)
    weather_result = "Weather.com says: It is " + weather_com_
    result['current_conditions']['text'].lower() + " and " + weather_com_
    result['current_conditions']['temperature'] + "degree celcius now in " +
    city_name

    tts(weather_result)
```

In the first line of the code, you import the `pywapi` module. Then you code the weather function, which takes `city_name` and `city_code` as arguments. You store the result in the `weather_com_result` variable. You then access the current conditions' text and temperature from the result received from the `pywapi.get_weather_from_weather_com()` function. You store the appropriate message for the user by assigning it to the `weather_result` variable, which is then spoken by the `tts()` function.

Finally, it is time to make the edits in `brain.py`. Edit the first line so that it imports the weather module you built:

```
from GreyMatter import tell_time, general_conversations, weather
```

Now, edit the `brain()` function's declaration:

```
def brain(name, speech_text, city_name, city_code):
```

And add the code snippet to detect a weather query in the `if/else` ladder:

```
elif check_message(['how','weather']) or check_message(['hows', 'weather']):
    weather.weather(city_name, city_code)
```

This concludes the construction of the weather module for Melissa's logical engine. Now you can ask questions such as, "How is the weather?" and "How is the weather today?" and Melissa will let you know!

Adding the weather feature was straightforward because it involved only simple use of the module, and you didn't have to write code to retrieve the weather information from `weather.com`. The next example also uses a module, but its implementation is interesting and will help you brainstorm about adding new features to Melissa and the procedure of implementing them.

# Define Artificial Intelligence!

In this example, you retrieve definitions of and information about particular keywords from Wikipedia. This will let you ask Melissa about anything that has an article on Wikipedia. For this command, you use a specific format: "Define subject."

---

■ **Note**   I would like to point out that for proper implementation of this type of functionality, a question like "Who is Tanay Pant?" should be synonymous with "Define Tanay Pant." This would be possible by implementing natural language processing (NLP). Many NLP-based tools are available for research and development work, such as Natural Language Processing Toolkit (NLTK). You may want to read up on this topic, but let me warn you that it is a vast field. Covering NLP is beyond the scope of this book.

---

Before you start building the module, install the `wikipedia` module via pip by entering the following command in the terminal:

```
$ pip install wikipedia
```

Create a file named define_subject.py in the GreyMatter folder, and enter the following code:

```python
import re

import wikipedia

from SenseCells.tts import tts

def define_subject(speech_text):
    words_of_message = speech_text.split()
    words_of_message.remove('define')
    cleaned_message = ' '.join(words_of_message)

    try:
        wiki_data = wikipedia.summary(cleaned_message, sentences=5)

        regEx = re.compile(r'([^\(]*)\(([^\)]*)\) *(.*)')
        m = regEx.match(wiki_data)
        while m:
            wiki_data = m.group(1) + m.group(2)
            m = regEx.match(wiki_data)

        wiki_data = wiki_data.replace("'", "")
        tts(wiki_data)
    except wikipedia.exceptions.DisambiguationError as e:
        tts('Can you please be more specific? You may choose something from the following.')
        print("Can you please be more specific? You may choose something from the following.; {0}".format(e))
```

This code imports the regular-expressions module named re and the wikipedia module that you installed via pip. The best approach to understand this code is working through a sample case.

Suppose the user gives the command "define tanay pant." It is passed as an argument to the define_subject() function. The string is split into an array of words, from which the word *define* is removed. This leaves you with the following two words in the array: *tanay* and *pant*. These two words form the subject that needs defining. This new array of words is rejoined and assigned to the cleaned_message variable. It then seeks a summary from wikipedia via the wikipedia.summary() function. You specify two arguments: the subject—that is, the cleaned_message variable—and the number of sentences the summary should contain.

The next statement consists of a regular-expression pattern match that removes anything in braces (braces inclusive) from the summary and then recombines the summary. You do this because braces will mess with the tts() function. It then removes all the apostrophe (') characters from the summary, because they also interfere with tts(). Finally, the result is spoken via the tts() function.

You include the except statement for the disambiguation error, because if the user asks for the definition of a subject like *hacker*, which has multiple meanings, a more specific subject must be defined. The list of specific subjects is shown as an output in the console.

Now, make the following changes in the brain.py file:

```
from GreyMatter import tell_time, general_conversations, weather, define_subject
```

Then, add the code snippet to recognize this feature from the user's command to the if/else ladder:

```
elif check_message(['define']):
    define_subject.define_subject(speech_text)
```

Running "define tanay pant" gives me the following speech output:

> *Tanay Pant is an Indian author, hacker, developer and tech enthusiast. He is best known for his work on "Learning Firefox OS Application Development" which was published by Packt. He is also an official representative of Mozilla. He has been listed in the about:credits of the Firefox web browser for his contributions to the different open source projects of the Mozilla Foundation.*

If you look at the page on Wikipedia (https://en.wikipedia.org/wiki/Tanay_Pant) from which it was retrieved, notice that "(born 28 September 1995)" has been removed from the output given by Melissa (see Figure 4-2).

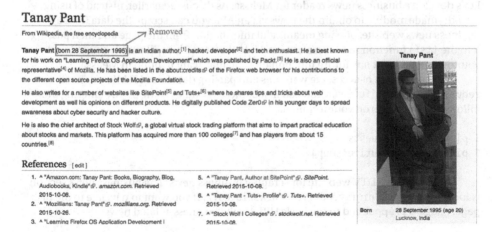

*Figure 4-2. Sample Wikipedia page*

Let's revise the workflow for the keyword-specific functionality, where some part of the speech is a keyword and the other part is information that needs to be used by the logic engine. The illustration in Figure 4-3 summarizes what you just learned in the form of a flowchart.

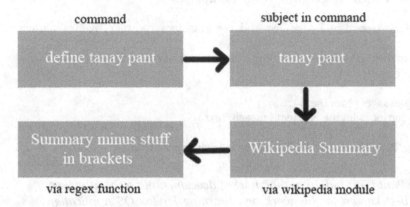

*Figure 4-3.* *Wikipedia information-retrieval workflow*

Melissa now has a functional definition system that lets you ask about a wide variety of subjects. This particular module has helped you to boost the functionality of your beloved virtual assistant.

# Read Me Some Business News!

Let's develop a business news reader for Melissa. As discussed earlier, instead of using a ready-made module to obtain the news via an API, you can scrape the data from a business news web site, parsing meaningful information from the page and then passing it to the tts() function so Melissa can read it to the user. This will enable you to build your own module for any future functionality you may want to build in Melissa.

For the purpose of accessing web sites and parsing data from HTML, you need requests and BeautifulSoup. You can install these modules using pip by entering the following commands on your terminal:

```
$ pip install requests
$ pip install beautifulsoup4
```

I selected the NDTV web site for scraping business news. The business news section is located at http://profit.ndtv.com/news/latest/. If you want to navigate to the next page, /page-2 is appended to the basic URL for the business-related news.

Studying the source reveals that the headlines are in <h2> tags. You may also notice that two unnecessary <h2> tags are present and that a brief summary of the headlines appears in <p> tags with intro set as the class. Now, having made the necessary observations for building the news reader, create a file named business_news_reader.py in the GreyMatter folder and enter the following code:

```
import requests
from bs4 import BeautifulSoup

from SenseCells.tts import tts

# NDTV News
fixed_url = 'http://profit.ndtv.com/news/latest/'
news_headlines_list = []
news_details_list = []

for i in range(1, 2):
    changing_slug = '/page-' + str(i)
    url = fixed_url + changing_slug
    r  = requests.get(url)
    data = r.text

    soup = BeautifulSoup(data, "html.parser")

    for news_headlines in soup.find_all('h2'):
        news_headlines_list.append(news_headlines.get_text())

    del news_headlines_list[-2:]

    for news_details in soup.find_all('p', 'intro'):
        news_details_list.append(news_details.get_text())

news_headlines_list_small = [element.lower().replace("(", "").replace(")",
"").replace("'", "") for element in news_headlines_list]
news_details_list_small = [element.lower().replace("(", "").replace(")",
"").replace("'", "") for element in news_details_list]

news_dictionary = dict(zip(news_headlines_list_small, news_details_list_small))

def news_reader():
    for key, value in news_dictionary.items():
        tts('Headline, ' + key)
        tts('News, ' + value)
```

First, you import the requests module as well as BeautifulSoup from the bs4 module. You assign the fixed URL that you found on the web site to the fixed_url variable. Then, you declare two empty arrays named news_headlines_list and news_details_list, which will hold the headlines and the news, respectively.

Next, you have to cycle through the different pages to scrape news from them. You do this by using a `for` loop and adding a dynamic `changing_slug` whose value depends on the iterations of the loop. This helps to create the URL, which you call using `requests.get()` to capture the page's HTML to the `data` variable.

You then call the `BeautifulSoup` parser, which iterates through all the `<h2>` tags and `<p>` tags with class `intro` to create two separate lists for headlines and news. You can remove the two unwanted `<h2>` tags by slicing the headline list appropriately.

The next step involves editing the contents of both lists to remove parentheses and quotes and make the text lowercase. Then you create a dictionary that stores the headlines and news as key/values pairs. Finally, you build the `news_reader()` function, which iterates through the items in the dictionary, and have Melissa speak the headlines and news via the `tts()` function.

---

■ **Note**    This approach is vulnerable to changes in the site's HTML. I used the data-mining approach to demonstrate how you can retrieve information if no other alternative is available. Using the official API or a web site's RSS feed will solve this problem.

---

The last step is adding the appropriate information to `brain.py`:

```
from GreyMatter import tell_time, general_conversations, weather,
define_subject, business_news_reader
```

Now add the corresponding code snippet to the `if`/`else` clause:

```
elif check_message(['business', 'news']):
    business_news_reader.news_reader()
```

Congratulations—you have just built a business news reader for Melissa! I am sure she is grateful. You can call this functionality by saying a command like, "Read me the business news!" or "Latest business news!" Let's revise the workflow you follow to obtain the news from the news web site and process it in such a way that it is appropriate to pass to `tts()`; see Figure 4-4.

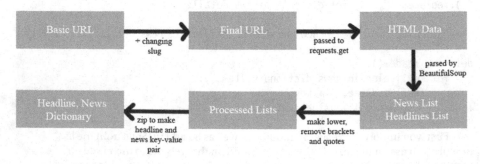

***Figure 4-4.*** *News-retrieval workflow*

Similarly, following this workflow, you can scrape data from any web site from which you want to retrieve data and build modules to retrieve other types of information from the Internet.

## Text-Controlled Virtual Assistant

Here is an exercise for you: try to modify Melissa's code such that instead of executing the main.py Python file in the usual way, you give the following command:

```
$ python main.py -t
```

As a result, Melissa will take text input instead of voice input. This will help you to interact with the virtual assistant if you cannot speak for some reason. This may come in handy if your microphone does not work or if you are in a public room where speaking a command to your computer might be awkward.

To see my implementation of the text-controlled virtual assistant flag, visit the Melissa-Core repository under the Melissa-AI organization and take a look at the main.py file. But I really encourage you to try to implement this yourself before opening Melissa's repository.

## Selenium and Automation

If you have had experience in quality assurance or automated testing, you can automate your daily web site testing with Melissa's help via Selenium. Even if you do not have any experience using Selenium, you should be pleased to know that you can do fun things with it. First, install selenium using pip:

```
$ pip install selenium
```

Now, create a file named open_firefox.py in the GreyMatter folder, and type the following code in it:

```
from selenium import webdriver

from SenseCells.tts import tts

def open_firefox():
    tts('Aye aye captain, opening Firefox')
    webdriver.Firefox()
```

On the first line, you import webdriver from the selenium module. Then you call the function webdriver.Firefox(), which opens the Firefox web browser. To implement it, open brain.py and make the following code changes/additions:

```
from GreyMatter import tell_time, general_conversations, weather,
define_subject, business_news_reader, open_firefox

    elif check_message(['open', 'firefox']):
        open_firefox.open_firefox()
```

This lets you open Firefox by giving commands like "Open Firefox."

You can do a host of pretty interesting things with Selenium. For example, you can automate logging in to a web site like Facebook so that when you give the command to Melissa, she opens a browser, opens the web site, and then logs in to the web site using your credentials.

Some Internet connections need the user to log in to proxy portals, also known as *captive portals*, to access the Internet. You can also automate this type of login. Let's write a module for performing this functionality.

Store the proxy username and password in the profile.yaml file:

```
proxy_username:
  Something
proxy_password:
  Something
```

I would like to point out that saving passwords in plaintext and in publicly accessible files is a very bad idea. You may wish to save the password in a database in an encrypted format. However, for the sake of simplicity, keep it in profile.yaml for now.

Make the following additions to main.py:

```
proxy_username = profile_data['proxy_username']
proxy_password = profile_data['proxy_password']
```

Edit the call to the brain() function in main() as follows (and also in the function declaration in brain.py):

```
brain(name, speech_text city_name, city_code, proxy_username, proxy_password)
```

Now, create a connect_proxy.py file in the GreyMatter folder, and type the following code in it:

```
from selenium import webdriver

from SenseCells.tts import tts

def connect_to_proxy(proxy_username, proxy_password):
    tts("Connecting to proxy server.")
    browser = webdriver.Firefox()
    browser.get('http://10.1.1.9:8090/httpclient.html')

    id_number = browser.find_element_by_name('username')
    password = browser.find_element_by_name('password')

    id_number.send_keys(proxy_username)
    password.send_keys(proxy_password)

    browser.find_element_by_name('btnSubmit').click()
```

As is evident from the code, the connect_to_proxy() function first opens the browser, then opens the proxy URL, and then searches for the username text field and the password field. Finally, it enters the username and the password in their respective fields and presses Enter to log in the user.

You can make a similar login-automation module for any web site by first making the changes in the URL, username, and password ID, class, or name (from the source code), and changing the username and password.

Make the final edits and additions in brain.py:

```
from GreyMatter import tell_time, general_conversations, weather,
define_subject, business_news_reader, open_firefox, connect_proxy
    elif check_message(['connect', 'proxy']):
        connect_proxy.connect_to_proxy(proxy_username, proxy_password)
```

Now, giving Melissa a command like "Connect to proxy server" will cause her to automatically connect you to the proxy server.

Remember that you should never store passwords in plaintext in publicly accessible files. As I mentioned, you may want to re-create the earlier instructions and store the username and password in a database using an encryption scheme such as SHA2.

# Time to Sleep, Melissa!

Let's write a short module to shut down the software and ask Melissa to sleep. Type the following code for sleep.py in the GreyMatter folder:

```
from SenseCells.tts import tts

def go_to_sleep():
    tts('Goodbye! Have a great day!')
    quit()
```

Time to make the edits in brain.py:

```
from GreyMatter import tell_time, general_conversations, weather,
define_subject, business_news_reader, open_firefox, connect_proxy, sleep

    elif check_message(['sleep']):
        sleep.go_to_sleep()
```

Now, saying something like "Time to sleep, Melissa!" or "Sleep!" will shut down the software and exit from the Python script.

You should commit your code after building a feature or completing the lessons in any chapter. This will help you to go back to your previous state (the last time you committed) if you accidently delete something or mess something up. You can go back to the previously committed state by entering the following command on your terminal:

```
$ git reset --hard
```

Trust me, this command has saved my life many more times than I can remember. I hope this will help you some day as well.

# Summary

In this chapter, you learned how to build applications that use the Internet to gather useful information and present to the user when requested. You used modules to retrieve weather information and definitions from Wikipedia. You also learned how to mine data from web sites to extract meaningful data when a third-party module is not available. You learned how to use Selenium to create some elementary features for Melissa. Finally, you created a small sleep module for Melissa that shuts down the virtual assistant.

In the next chapter, you build a music player module for Melissa, which allows you to ask Melissa to randomly play any music file as well as search for music and play a file.

# CHAPTER 5

■ ■ ■

# Developing a Music Player for Melissa

This chapter covers the details of building a music player for Melissa that lets you select all the MP3 files in a given directory and play them using a command-line music utility. Melissa will be able to play music randomly as well as play a specific music file from a list of files when asked.

To build this functionality for Melissa, you must first select command-line players for OS X and Linux so that after appropriate handling by the logic engine, it can pass the name of a song or a list of songs to the music player via the os.system() function. Let's discuss the command-line music players suitable for this module before moving forward to building the module.

## OS X Music Player

Apple's OS X has a built-in command-line music player utility called afplay that helps you play music from the command line without installing anything. To check whether you have afplay installed on your system, open the terminal and enter the following command:

```
$ afplay
```

This should give you the following output:

```
Tanays-MacBook-Air:~ tanay$ afplay

    Audio File Play
    Version: 2.0
    Copyright 2003-2013, Apple Inc. All Rights Reserved.
    Specify -h (-help) for command options

Usage:
afplay [option...] audio_file
```

© Tanay Pant 2016

T. Pant, *Building a Virtual Assistant for Raspberry Pi*, DOI 10.1007/978-1-4842-2167-9_5

```
Options: (may appear before or after arguments)
  {-v | --volume} VOLUME
    set the volume for playback of the file
  {-h | --help}
    print help
  { --leaks}
    run leaks analysis
  {-t | --time} TIME
    play for TIME seconds
  {-r | --rate} RATE
    play at playback rate
  {-q | --rQuality} QUALITY
    set the quality used for rate-scaled playback (default is 0 - low
quality, 1 - high quality)
  {-d | --debug}
    debug print output
```

## Linux Music Player

For Linux, I recommend installing mpg123 from its official web site, www.mpg123.de.
It is a sleek, decent music player utility like afplay. You can run mpg123 by entering the
following command in your console:

```
$ mpg123 'something.mp3'
```

Doing so plays the music file you specify.

## Module Workflow

First you need to find all the MP3 files present in the path specified by the user as well
as in its subdirectories, and make a list. Next, you either choose a random music file to
play and send it to the music player or search the music list to find the song the user
requested. Then, you fingerprint the OS to determine which music player to summon
(see Figure 5-1).

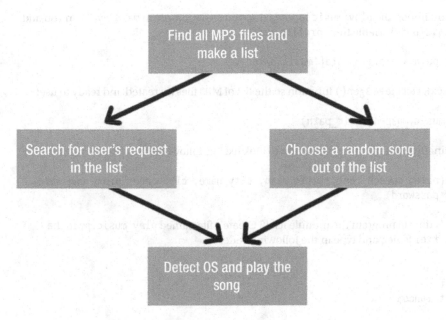

**Figure 5-1.** *Music player workflow*

The work specified by the first block on the flowchart, "Find all MP3 files and make a list," is accomplished by using a function named mp3gen(). The second block's work, "Search for user's request in the list," is done by a function named play_specific_music(). The third block's work, "Choose a random song out of the list," is done by a function named play_random(). And, finally, the last block's work, "Detect OS and play the song," is done by the music_player() function.

## Building the Music Module

Let's get to the task of programming the flowchart just discussed. First, add the path of the folder where the music resides in the profile.yaml file:

```
music_path:
```

•

Next you need to extract the information about the path of the music files in main.py so that it can pass that information to brain.py. Also, you need to make a couple of additions:

```
from GreyMatter import play_music
```

˙      You import the play_music module that you create shortly in main.py. Then you add the music_path variable from profile.yaml:

```
music_path = profile_data['music_path']
```

Next, call the mp3gen() function so the list of MP3 files is created and ready to use:

```
play_music.mp3gen(music_path)
```

Finally, edit the brain() function to look like the following:

```
brain(name, speech_text, music_path, city_name, city_code, proxy_username, proxy_password)
```

It's time to program the module itself! Create a file named play_music.py in the GreyMatter folder, and type in the following code:

```python
import os
import sys
import random

from SenseCells.tts import tts

def mp3gen(music_path):
    """
    This function finds all the MP3 files in a folder and its subfolders and
    returns a list:
    """
    music_list = []
    for root, dirs, files in os.walk(music_path):
        for filename in files:
            if os.path.splitext(filename)[1] == ".mp3":
                music_list.append(os.path.join(root, filename.lower()))
    return music_list

def music_player(file_name):
    """
    This function takes the name of a music file as an argument and plays it
    depending on the OS.
    """
    if sys.platform == 'darwin':
        player = "afplay '" + file_name + "'"
        return os.system(player)
    elif sys.platform == 'linux2' or sys.platform == 'linux':
        player = "mpg123 '" + file_name + "'"
        return os.system(player)
```

```python
def play_random(music_path):
    try:
        music_listing = mp3gen(music_path)
        music_playing = random.choice(music_listing)
        tts("Now playing: " + music_playing)
        music_player(music_playing)
    except IndexError as e:
        tts('No music files found.')
        print("No music files found: {0}".format(e))

def play_specific_music(speech_text, music_path):
    words_of_message = speech_text.split()
    words_of_message.remove('play')
    cleaned_message = ' '.join(words_of_message)
    music_listing = mp3gen(music_path)

    for i in range(0, len(music_listing)):
        if cleaned_message in music_listing[i]:
            music_player(music_listing[i])
```

Starting from the beginning, you import the built-in os, sys, and random. Next comes the mp3gen() function. In this function, you pass music_path as an argument. You declare an empty list to hold the array of music file names. You then iterate through the files, folders, and subfolders using the os.walk() function to find all files with the .mp3 extension. When it finds the required files, it stores the names of the files along with their complete path address to the music_list variable. The function returns music_list as a list (array).

The music_player() function is written to play the music files after detecting the user's OS. The function takes file_name as an argument. Similar to what you did while building the tts() function earlier, you use the sys.platform() function to detect whether the OS is OS X or Linux. Accordingly, you create a variable named player in which you concatenate the player along with the name of the music file to play; you use either the afplay player or the mpg123 player. This player variable acts as a command that is called using the os.system() command.

Next comes the play_random() function, where you create the list of all MP3 files present using the mp3gen() function. This function takes music_path as an argument. Then you create a variable named music_playing that stores the name of a particular music file by using the random.choice() function, which operates on the music_listing list. You then pass the name of the music file stored in music_playing to the music_player() function, which plays the music. You use a try/except clause here because there may be a case when there are no MP3 files present in the music_path; this gives an IndexError, which speaks the message "No music files found."

Finally, the play_specific_music() function takes speech_text and music_path as arguments. You implement the same functionality here as in the define_subject module. So, you split speech_text to create an array of words. You then remove the *play* keyword from the array, and whatever remains, however improbable it may be, must be the name of the music file the user wants to search for. You combine the words of the array again and iterate through music_list to find a match with the name of the song the user specified. If a match is found, the music is played using the music_player() function.

Now it's time to edit brain.py; make the following edits and additions. The first change is to the import statement:

```
from GreyMatter import define_subject, tell_time, general_conversations,
play_music, weather, connect_proxy, open_firefox, sleep, business_news_reader
```

Edit the declaration of the brain() function to make it look like the following:

```
def brain(name, speech_text, music_path, city_name, city_code, proxy_
username, proxy_password):
```

The last step is to add a code snippet to call the two functions in the file's logic-handling if/else clause:

```
elif check_message(['play', 'music']) or check_message(['music']):
    play_music.play_random(music_path)

elif check_message(['play']):
    play_music.play_specific_music(speech_text, music_path)
```

So far, so good! Note that you purposely put the play_random() function first, because it recognizes the call made to it via the *play* and *music* keywords. If you put just the *play* keyword first, as in the play_specific_music() function, then even if you want to hear a random track, the module would split the query and completely mess it up, resulting in "track not found" and hence an error. The first clause in an if/else ladder takes the first priority.

## Play Party Mix!

This is an exercise for you: create another function in the play_music module that makes a list of all the MP3 files, shuffles them, and then plays them one by one. This feature should be invoked when keywords such as *party* and *mix* or *party* and *time* are present.

I encourage you to try to implement this feature yourself before looking at the following solution. Doing so will improve your understanding of this software and help you scale the software in the future.

First, make a function named play_shuffle() in play_music.py. Type in the following code:

```
def play_shuffle(music_path):
    try:
        music_listing = mp3gen(music_path)
        random.shuffle(music_listing)
        for i in range(0, len(music_listing)):
            music_player(music_listing[i])
    except IndexError as e:
        tts('No music files found.')
        print("No music files found: {0}".format(e))
```

This function accepts the music_path variable and makes a list of all the MP3 files that are present. It then shuffles the music list using the random.shuffle() function. Doing so ensures that the order in which the songs are played is different each time. Now you iterate through the shuffled list and play the music files one by one. If there is an IndexError exception, it passes on the message "No music files found."

Make the following changes in brain.py and add this code snippet:

```
elif check_message(['party', 'time']) or check_message(['party', 'mix']):
    play_music.play_shuffle(music_path)
```

This ensures that the party mix is called when the user says something like, "It's party time!" or "Party mix!" Note that you must add this code snippet above the time module's code snippet. If you don't, then if you say "It's party time!," the software will detect the *time* keyword, and the logical engine will transfer control of the program to the time module.

# Summary

In this chapter, you learned how to build music-player functionality for Melissa, and you learned to find all the MP3 files in a folder and its subfolders, make a list of them, play them randomly, search for specific music, and create a party mix. Did I forget to mention that you made all of this voice controlled?

In the next chapter, you build a voice-controlled note-taking application that lets you save notes using your voice and retrieve them later.

# CHAPTER 6

■ ■ ■

# Developing a Note-Taking Application

A virtual assistant would not be truly useful if it didn't help with tasks such as writing notes for you so you don't have to pick up a pen or type your creative thoughts on the keyboard. This chapter covers the steps to build a voice-controlled, note-taking application using Python code and a SQLite3 database. This application lets the reader save any message in the database with the help of the *note* keyword and retrieve the notes later by asking Melissa to do so.

The application uses a SQLite database to save the notes. Python by default installs SQLite3, so don't worry about having to install anything in this chapter.

This chapter's very simple voice-controlled, note-taking application is invoked if it finds the *note* keyword. It follows the same message-extraction technology you used to build the define_subject module and the music-search functionality in the play_music module.

## Design Workflow

In this application, you search speech_text for the keyword *note*. It then removes the word *note* from speech_text to extract the note. You rejoin the message and insert the note into the database along with the date. When it is saved successfully, you see an alert (see Figure 6-1).

© Tanay Pant 2016

T. Pant, *Building a Virtual Assistant for Raspberry Pi*, DOI 10.1007/978-1-4842-2167-9_6

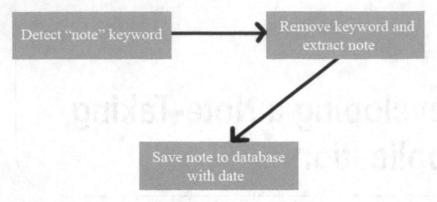

**Figure 6-1.** *Note-saving workflow*

# Designing the Database

First things first: you need to design a SQLite3 database for storing notes. You have to create the database in the same format you used for the `profile.yaml` file (to protect the user's private data, which may be stored in the database). Give the database the name `memory.db.default`. Remember to make the following addition to the `.gitignore` file:

```
memory.db
```

To successfully run the module and work on the database, the user must enter the following command in the terminal:

```
$ cp memory.db.default memory.db
```

Enter the following command in your terminal to set up your database. This opens the database in the SQLite3 prompt:

```
$ sqlite3 memory.db
```

Now you have to create a table named `notes` containing two columns named `notes` and `notes_date`. The datatype for both columns is TEXT, and the fields cannot be null. So, enter the following command at the `sqlite` prompt:

```
sqlite> CREATE TABLE notes(
   ...> notes TEXT NOT NULL,
   ...> notes_date TEXT NOT NULL
   ...> );
```

You can check whether you are working on the correct database by entering the following command at the `sqlite` prompt:

```
sqlite> .databases
```

If the database you intend to work on in assigned the name `main`, then you have nothing to worry about. You can also consult the schema of your SQL database by entering the following command at the `sqlite` prompt:

```
sqlite> .schema
```

Here is the output of the commands I ran in my terminal. You should see something similar:

```
Tanays-MacBook-Air:Melissa-Core-master tanay$ sqlite3 memory.db
SQLite version 3.8.10.2 2015-05-20 18:17:19
Enter ".help" for usage hints.
sqlite> .databases
seq  name             file
---  ---------------  -------------------------------------------------------
0    main             /Users/tanay/Desktop/Melissa-Core-master/memory.db
sqlite> .schema
CREATE TABLE notes(
notes TEXT NOT NULL,
notes_date TEXT NOT NULL
);
sqlite> .exit
Tanays-MacBook-Air:Melissa-Core-master tanay$
```

# Inner Workings of the Virtual Assistant

Before moving on to develop the note-taking module, I want to revisit a topic from Chapter 1. Adding a database makes an important modification in Melissa's workflow. Now the virtual assistant has a memory like a human and can store important information that may be useful for the user or the functionality of modules you develop later. The workflow now looks as shown in Figure 6-2.

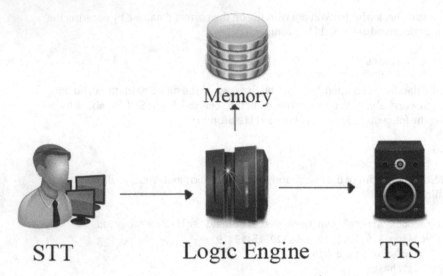

*Figure 6-2.* *Melissa's new structure*

The virtual assistant now has a feature to process information in the logical engine and pass it on to the memory if instructed to do so by the Python code. This new addition makes Melissa smarter and even more useful.

# Building the Note-Taking Module

It's time to jump into the task of writing the Python code for the note-taking module. Create a file named notes.py in the GreyMatter folder, and type the following code in it:

```python
import sqlite3
from datetime import datetime

from SenseCells.tts import tts

def note_something(speech_text):
    conn = sqlite3.connect('memory.db')
    words_of_message = speech_text.split()
    words_of_message.remove('note')
    cleaned_message = ' '.join(words_of_message)

    conn.execute("INSERT INTO notes (notes, notes_date) VALUES (?, ?)",
    (cleaned_message, datetime.strftime(datetime.now(), '%d-%m-%Y')))
    conn.commit()
    conn.close()

    tts('Your note has been saved.')
```

The starting line imports the built-in `sqlite3` and `datetime` modules. You then define a function named `note_something()` that takes `speech_text` as an argument. In the function, you establish a connection to the `memory.db` database using the `sqlite3.connect()` function.

You then split `speech_text`, remove the *note* keyword, extract the note by joining the rest of the words again, and assign the note to the `cleaned_message` variable. You remove the *note* keyword so that Melissa stores only the message that the user intends to store—*note* is just a command and not part of the user's message. Using the SQLite3 `execute()` function, you enter the SQL statement for inserting `cleaned_message` in the notes column and the date in the `notes_date` column of the `notes` tables. You obtain the date using the `datetime.now()` function and format it using the `datetime.strftime()` function. You then commit the changes made to the database using the `commit()` function and close the database using the SQLite3 `close()` function. Finally, you give oral feedback to the user, telling them that their note has been successfully saved.

You now have to make the changes to the `brain.py` file. Make the following edits and additions:

```
from GreyMatter import notes, define_subject, tell_time, general_
conversations, play_music, weather, connect_proxy, open_firefox, sleep,
business_news_reader
```

Next, add the appropriate code snippet to the `if/else` ladder in Melissa's logical engine:

```
elif check_message(['note']):
    notes.note_something(speech_text)
```

Congratulations—you have built note-taking functionality for your virtual assistant, Melissa! Now you can ask Melissa to jot down any important thoughts that come to mind. This feature makes Melissa even more useful for daily use. You can save your thoughts by giving a command such as, "Note remember to go to college!" This will save the note "Remember to go to college!"

You should try this activity with Melissa to save a note. To check whether your note has been successfully saved, open the terminal and open the `memory.db` database at the `sqlite` prompt. Enter the following command:

```
sqlite> select * from notes;
```

This shows you all the data that has been saved in the `notes` table. Entering it in my terminal shows me this output:

```
Tanays-MacBook-Air:Melissa-Core-master tanay$ sqlite3 memory.db
SQLite version 3.8.10.2 2015-05-20 18:17:19
Enter ".help" for usage hints.
sqlite> select * from notes;
remember to go to college! | 14-01-2016
sqlite> .exit
Tanays-MacBook-Air:Melissa-Core-master tanay$
```

# Building a Note-Dictating Module

It is okay for a developer to look at the data stored in the database by accessing the SQLite3 data via the command prompt. However, it would not convenient for the user to have to revisit the note data in this way. It would be better if Melissa could dictate previously saved notes to you. Let's build a module for retrieving old notes.

Create a new function named show_all_notes() in the notes.py file, and type the following code in it:

```python
def show_all_notes():
    conn = sqlite3.connect('memory.db')
    tts('Your notes are as follows:')

    cursor = conn.execute("SELECT notes FROM notes")

    for row in cursor:
        tts(row[0])

    conn.close()
```

The first line in the show_all_notes() function establishes the connection to your database. You then give the message that the notes are being dictated. You execute the same SQL statement that you entered in the terminal earlier to receive the data, iterate through the records to select the notes, and pass them to the tts() function. Finally, you close the connection to the database.

Add the appropriate code snippet to brain.py:

```python
elif check_message(['all', 'notes']) or check_message(['notes']):
    notes.show_all_notes()
```

Now commands such as "Show me all the notes!" or "Notes" will cause Melissa to read all the notes to you. You use or to ensure that notes.show_all_notes() is called if the user's command contains both the *all* and *notes* keywords (check_message() returns True only if both *all* and *notes* are present) and even if it contains just the *notes* keyword.

# Exercises

Here are some features for you to implement in the note-taking application to increase its usefulness. First, it would be great if a command like "Show today's notes!" would display the notes you saved today. This would help you easily review all the notes you saved at the end of the day.

Also, it would make sense to have a command such as "Delete all notes!" or "Delete today's notes!" This would help you prune unneeded notes from the database so that Melissa does not dictate notes you've already taken care of.

# Summary

In this chapter, you learned to develop a note-taking application that extracts a note from a user's command and saves it to a SQLite3 database. You then built a note-retrieval system that lets you retrieve the saved notes by giving another command.

In the next chapter, you learn how to develop a voice-controlled interface to access Twitter with Melissa's help. You also study how to build a voice-controlled image uploader that can upload images to Imgur (an image-upload web site).

# CHAPTER 7

■ ■ ■

# Building a Voice-Controlled Interface for Twitter and Imgur

In this chapter, you learn how to post status updates (tweets) on Twitter by using your voice. You also learn how you can upload images on Imgur by asking Melissa to do so, and save the URLs of the uploaded images in a database so they can be retrieved for future reference.

Twitter is the most famous microblogging platform and has a huge number of users. People love to post information about how they feel, interact with others, and express their views on different subjects. Wouldn't it be cool if you could tell Melissa what you want your new tweet to be, and she could post it for you? That would be a really helpful feature!

## Building the Twitter Module

Let's start by installing a third-party module named tweepy from pip. To install tweepy, enter the following command in your terminal:

```
$ pip install tweepy
```

Now you need to get the Twitter consumer key, consumer key secret, access token, and access token secret to authenticate the Twitter user via OAuth 2. To register an application on Twitter, go to apps.twitter.com and click Create New App. Enter the appropriate details for your application, and create the application. This generates the keys and access tokens required to authenticate your application.

© Tanay Pant 2016
T. Pant, *Building a Virtual Assistant for Raspberry Pi*, DOI 10.1007/978-1-4842-2167-9_7

After going to your application page, click the Keys and Access Tokens tab, which displays all the required information to create your module. Figure 7-1 shows my consumer keys.

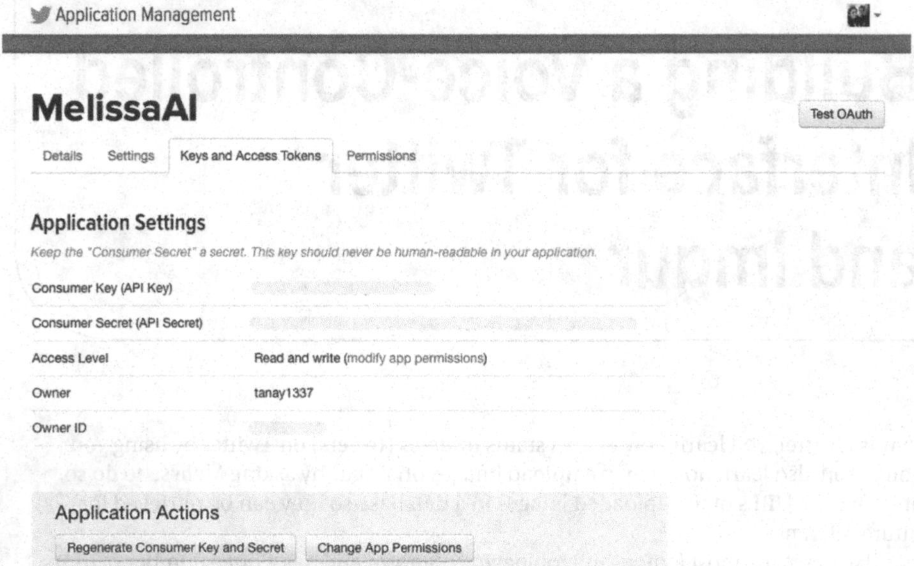

*Figure 7-1.* *Twitter application management*

Scrolling down the same page, you can get the access tokens for your Twitter application. Click the button to generate the access tokens. Figure 7-2 shows the access tokens after I clicked the button.

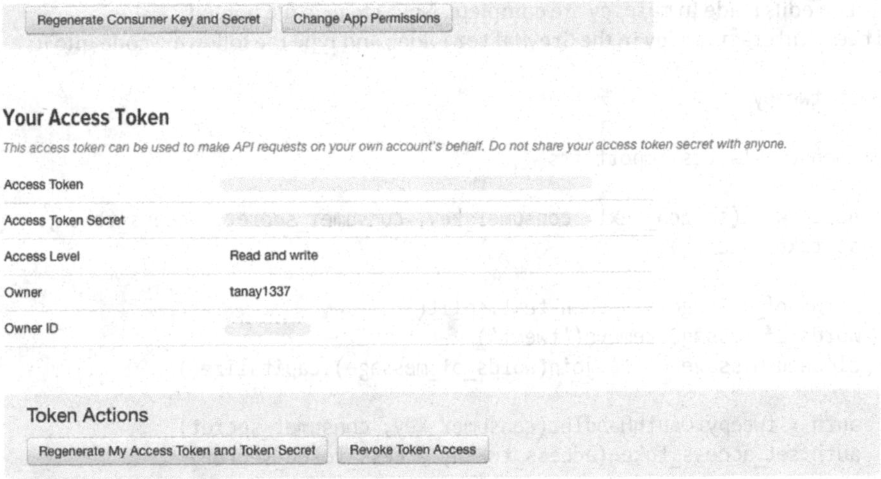

*Figure 7-2. Getting your access token*

Make the following additions to profile.yaml.default:

```
twitter:
  access_token:
    Something
  access_token_secret:
    Something
  consumer_key:
    Something
  consumer_secret:
    Something
```

Enter the correct values instead of Something in your profile.yaml file. You now need to extract these keys from the YAML file to your program; to do so, make the following changes in main.py:

```
access_token = profile_data['twitter']['access_token']
access_token_secret = profile_data['twitter']['access_token_secret']
consumer_key = profile_data['twitter']['consumer_key']
consumer_secret = profile_data['twitter']['consumer_secret']
```

Next you need to pass all these values to the brain() function by editing the call made to it in main.py:

```
brain(name, speech_text, city_name, city_code, proxy_username, proxy_
password, consumer_key, consumer_secret, access_token, access_token_secret)
```

The edits made in main.py are complete. Now, create a file named twitter_interaction.py in the GreyMatter folder, and type the following code into it:

```python
import tweepy

from SenseCells.tts import tts

def post_tweet(speech_text, consumer_key, consumer_secret, access_token,
access_token_secret):

    words_of_message = speech_text.split()
    words_of_message.remove('tweet')
    cleaned_message = ' '.join(words_of_message).capitalize()

    auth = tweepy.OAuthHandler(consumer_key, consumer_secret)
    auth.set_access_token(access_token, access_token_secret)

    api = tweepy.API(auth)
    api.update_status(status=cleaned_message)

    tts('Your tweet has been posted')
```

The first line imports the tweepy module. You define a function named post_tweet() that takes the following arguments: speech_text, consumer_key, consumer_secret, access_token, and access_token_secret. You then split speech_text, remove the *tweet* keyword from it, and rejoin it to extract the tweet text. Next, the tweepy.OAuthHandler() handler is called; it takes consumer_key and consumer_token as arguments. You set the access tokens using the set_access_token() function and pass the access_token and access_token_secret arguments. The tweepy API is called using the tweepy.api() function. Finally, you call the update_status() function, which takes status as an argument and posts it. On successfully completing this task, Melissa sends a voice message that "Your tweet has been posted!"

Now, make the following edits in brain.py:

```python
from GreyMatter import notes, define_subject, tell_time, general_
conversations, play_music, weather, connect_proxy, open_firefox, sleep,
business_news_reader, twitter_interaction
```

And finally, add the appropriate code snippet to the if/else ladder:

```python
elif check_message(['tweet']):
    twitter_interaction.post_tweet(speech_text, consumer_key,
    consumer_secret, access_token, access_token_secret)
```

This ensures that if you pass a command such as "Tweet hello twitter greetings from melissa," the application will remove the *tweet* keyword and post your message on Twitter. Figure 7-3 shows my Twitter wall after I gave Melissa this command.

*Figure 7-3.* *Tweet posted on Twitter*

This code works like a charm. Also notice that it has capitalized "Hello" because I used the `capitalize()` function on the `cleaned_message()` variable.

# Exercises

Here are a few tasks for you to accomplish by yourself in order to extend the functionality of Melissa's Twitter module and make it more useful. First, add a function in the `twitter_interaction` file to search for a person's username and dictate the details about them. Another interesting feature you can add is to have Twitter search for tweets on a particular topic or subject. You should also be able to hear your own tweets as well as the tweets of any other user. And a Twitter application is never complete in the truest sense if it cannot send direct messages to your followers.

Adding these functionalities to the `twitter_interaction.py` file will improve your understanding of the `tweepy` module and let you voice-control Twitter like a pro. If you successfully implement these features, be sure to send a pull request to the `Melissa-AI/Melissa-Core` repository on GitHub to get your code merged into the official repository.

# Building the Imgur Module

Imgur is an image-upload web site that provides an API and a Python module for uploading images and sharing them with your friends. Uploading images is a boring task—you have to click the Upload File button, select the file, and then click the Upload button. Drag-and-drop uploads certainly make the task easier, but you still have to find the file and use the mouse to perform the drag-and-drop operation.

It would definitely make things easier for you if Melissa could do all this for you, wouldn't it? Also, it would be even better if the URLs of the images you upload to Imgur are stored in a database, along with the dates. Let's build a module that can perform these tasks and simplify life for you!

First you have to install the imgurpython module, which acts as an interface to Imgur's REST API. To install imgurpython via pip, enter the following command in your terminal:

```
$ pip install imgurpython
```

Now you need the client_id and client_secret keys from Imgur. For this, create an account on Imgur, and request access from the registration form for your application. After creating the application, view your registered applications at https://imgur.com/account/settings/apps. Figure 7-4 shows the applications I have registered.

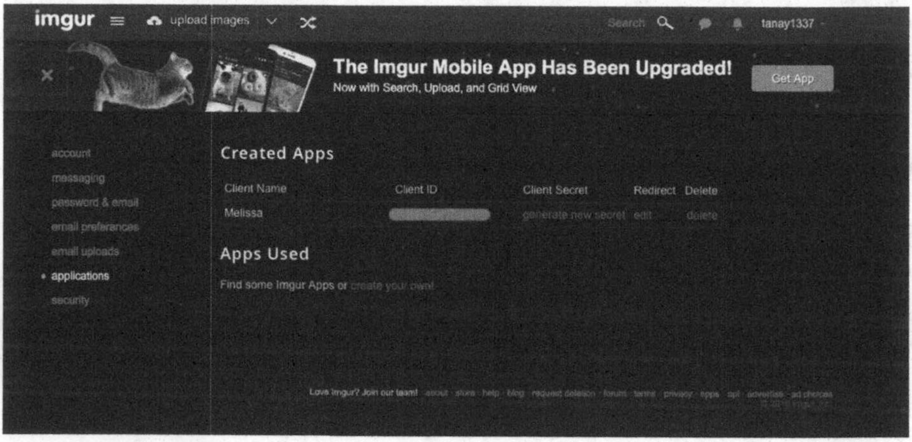

**Figure 7-4.** *Obtaining the Imgur client ID and secret*

After getting the keys, make the following additions to your profile.yaml.default file:

```
images_path:
  .
imgur:
  client_id:
    Something
  client_secret:
    Something
```

images_path has functionality similar to the music_path you used earlier, and the implementation of Imgur is similar to what you did with Twitter. Be sure to make the corresponding changes in the profile.yaml file.

## Creating the Tables in the Database

Now you need to create the appropriate tables in the database, to save the data. Open the memory.db database by entering the following command in the terminal:

```
$ sqlite3 memory.db
```

Enter the following SQL statement at the sqlite prompt to create the table:

```
sqlite> CREATE TABLE image_uploads(
   ...> filename TEXT NOT NULL,
   ...> url TEXT NOT NULL,
   ...> upload_date TEXT NOT NULL
   ...> );
```

You can now check the schema of the database by entering the following command:

```
Sqlite> .schema
```

Here is the output that I received on entering the previous command:

```
Tanays-MacBook-Air:Melissa-Core-master tanay$ sqlite3 memory.db
SQLite version 3.8.10.2 2015-05-20 18:17:19
Enter ".help" for usage hints.
sqlite> .schema
CREATE TABLE notes(
notes TEXT NOT NULL,
notes_date TEXT NOT NULL
);
CREATE TABLE image_uploads(
filename TEXT NOT NULL,
url TEXT NOT NULL,
upload_date TEXT NOT NULL
);
sqlite> .exit
Tanays-MacBook-Air:Melissa-Core-master tanay$
```

You need to extract this information in main.py and make the appropriate additions and edits. First, edit the import statement to make it look like this:

```
from GreyMatter import play_music, imgur_handler
```

Then extract the information from the YAML file:

```
images_path = profile_data['images_path']
client_id = profile_data['imgur']['client_id']
client_secret = profile_data['imgur']['client_secret']
```

Now, add the code to create the list of all the images in images_path and edit the call to the brain() function to pass the new arguments that are needed by your module:

```
imgur_handler.img_list_gen(images_path)
brain(name, speech_text, music_path, city_name, city_code, proxy_
username, proxy_password, consumer_key, consumer_secret, access_token,
access_token_secret, client_id, client_secret, images_path)
```

Note that the module and the function you add here have not yet been constructed.

Create a file named imgur_handler.py in the GreyMatter folder, and type the following code in it:

```
import os
import sqlite3
from datetime import datetime

from imgurpython import ImgurClient

from SenseCells.tts import tts

def img_list_gen(images_path):

    image_list = []
    for root, dirs, files in os.walk(images_path):
        for filename in files:
            if os.path.splitext(filename)[1] == ".tiff" or os.path.
            splitext(filename)[1] == ".png" or os.path.splitext(filename)[1]
            == ".gif" or os.path.splitext(filename)[1] == ".jpg":
                image_list.append(os.path.join(root, filename.lower()))
    return image_list

def image_uploader(speech_text, client_id, client_secret, images_path):

    words_of_message = speech_text.split()
    words_of_message.remove('upload')
    cleaned_message = ' '.join(words_of_message)
    image_listing = img_list_gen(images_path)

    client = ImgurClient(client_id, client_secret)

    for i in range(0, len(image_listing)):
        if cleaned_message in image_listing[i]:
            result = client.upload_from_path(image_listing[i], config=None,
            anon=True)
```

```
    conn = sqlite3.connect('memory.db')
    conn.execute("INSERT INTO image_uploads (filename, url, upload_date)
    VALUES (?, ?, ?)", (image_listing[i], result['link'], datetime.
    strftime(datetime.now(), '%d-%m-%Y')))
    conn.commit()
    conn.close()

    print result['link']
    tts('Your image has been uploaded')

def show_all_uploads():
    conn = sqlite3.connect('memory.db')

    cursor = conn.execute("SELECT * FROM image_uploads")

    for row in cursor:
        print(row[0] + ': (' + row[1] + ') on ' + row[2])

    tts('Requested data has been printed on your terminal')

    conn.close()
```

That's a lot of code! Let's go through it line by line. You need the sqlite3 and datetime modules to store the URLs where the images are uploaded and to get the date when the image was uploaded, respectively.

You import ImgurClient from the Imgur client to help you upload the images to Imgur. The img_list_gen() function takes images_path as an argument and searches for all the PNGs, GIFs, JPGs, and TIFFs, in the current folder and its subfolders. This code is essentially the same as the mp3gen() function you built in Chapter 5. It returns a list of all the files found with the extensions specified.

Next comes the image_uploader() function, which takes speech_text, client_id, client_secret, and images_path as arguments. Just as in the define_subject module, it splits the words, removes the keyword *upload*, and extracts the name of the image to be uploaded. You then pass client_id and client_secret to ImgurClient to authenticate.

Now you search for the specified image file among the list of images. If a match is found, you upload it to Imgur using the client.upload_from_path() function. This function takes the path of the image to be uploaded as an argument. You save the output of this function to a variable named result and save the image details to the database.

You use the SQL statement to insert the image values into the corresponding columns of the table. Note that result['link'] stores the link where the image was uploaded on Imgur. The result dictionary also stores a host of other information you may wish to save in the database or look at. You then commit the changes made to the database and close the connection to it. You also display the link where the image has been uploaded on the terminal, and Melissa gives the message that the image has been uploaded.

The last function is the show_all_uploads() function, which retrieves all the stored entries in the image_uploads table and displays the stored information in a formatted fashion on the terminal. Melissa gives a message that the requested data has been printed. This function allows you to view images that have been previously uploaded to Imgur by showing their URLs.

Finally, you need to make these appropriate edits and additions to `brain.py`:

```
from GreyMatter import notes, define_subject, tell_time, general_
conversations, play_music, weather, connect_proxy, open_firefox, sleep,
business_news_reader, twitter_interaction, imgur_handler
```

Edit the `brain()` function's arguments:

```
def brain(name, speech_text, music_path, city_name, city_code, proxy_
username, proxy_password, consumer_key, consumer_secret, access_token,
access_token_secret, client_id, client_secret, images_path):
```

And add this code snippet to the `if`/`else` ladder:

```
elif check_message(['upload']):
    imgur_handler.image_uploader(speech_text, client_id, client_secret,
    images_path)
```

```
elif check_message(['all', 'uploads']) or check_message(['all',
'images']) or check_message(['uploads']):
    imgur_handler.show_all_uploads()
```

You can now upload a file using a command such as "Upload hello." The application will upload an image filed named `hello` or that has `hello` in its name. You can see all the upload entries saved in the database by issuing a command such as "Show all uploads!" or "Show all images!" You can also check the entries in the database by entering the following at the `sqlite` prompt:

```
sqlite> select * from image_uploads;
```

The following output shows what I received on my terminal after entering the previous command. It displays the name of an image I uploaded (one of the illustrations from this book), the URL where it was uploaded, and the date on which it was uploaded:

```
Last login: Fri Jan 15 11:03:29 on ttys000
Tanays-MacBook-Air:~ tanay$ cd Desktop/Melissa-Core-master/
Tanays-MacBook-Air:Melissa-Core-master tanay$ sqlite3 memory.db
SQLite version 3.8.10.2 2015-05-20 18:17:19
Enter ".help" for usage hints.
sqlite> select * from image_uploads;
./news reader workflow.jpg|http://i.imgur.com/fIywson.jpg|15-01-2016
sqlite> .exit
Tanays-MacBook-Air:Melissa-Core-master tanay$
```

This shows the data in all the columns of the table separated by a pipe character (|). You can also see the formatted version of the data by saying the command "Show all uploads!" or "Show all images!"

# Summary

In this chapter, you learned how to tweet and use Twitter using your voice-controlled virtual assistant, Melissa. You also learned to build an image-finding and -uploading facility using the `imgurpython` module. In addition, you can retrieve the list of images you've uploaded, along with the date on which they were uploaded and the URL where they were uploaded, using a SQLite3 database—also known as Melissa's memory.

In the next chapter, you build a web interface for Melissa by using some open source JavaScript libraries to record the audio via the user's web browser and saving the file to a `.wav` format. This is sent to a Python program that sends the WAV file to the Google Speech Recognition API for recognition.

# CHAPTER 8

■ ■ ■

# Building a Web Interface for Melissa

In this chapter, you build a web interface for Melissa by using some open source JavaScript libraries to record the audio using the user's web browser and saving the file to a .wav format. The file is sent to a Python program, which sends it to the Google Speech Recognition API for recognition.

Accessing Melissa through your terminal may seem intimidating to users who are not used to working on the command line. Such an interface doesn't work for many people who are not developers. Your current interface is good for research and development purposes, but it is not a user-facing product. Building a web interface for operating Melissa will help. It isn't the best workflow for operating a virtual assistant like Melissa via the Web, but its simplicity guarantees that you can understand what needs to be done; after grasping the basic concepts, you can improve it.

## Operating Workflow

You can build a web interface with the help of Python's web.py. The user opens the web site to access Melissa's web interface, clicks a button, starts speaking, and then clicks the same button to stop the recording. Then the user clicks the Save button to save the .wav file. They upload the WAV file through the web form, which sends it to the Python server and responds accordingly (see Figure 8-1).

© Tanay Pant 2016
T. Pant, *Building a Virtual Assistant for Raspberry Pi*, DOI 10.1007/978-1-4842-2167-9_8

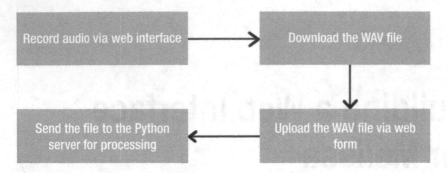

***Figure 8-1.*** *Operating workflow*

---

■ **Note**   The audio is recorded on the client side and must be imported from the browser into the system on the server. For this reason, a web interface is not a trivial addition to Melissa.

---

I would like to mention again that this is not the best approach to build a web interface. You implement it in this chapter for the sake of simplicity.

---

■ **Note**   Later in the chapter, I will invite you to interact with this application and improve it, as opposed to just reading and putting down the book. The web interface is a beta feature, which gives you a chance to work on something unique.

---

# Building the Web Interface

First you need to install Python's web module by entering the following command in the terminal:

```
$ pip install web.py
```

Now, create a new file called web-gateway.py in the root folder of your repository. This is the starting point of the Python server for serving your web application and so forth. Type the following code in the file:

```
import os
import yaml
import web

from GreyMatter.SenseCells.tts import tts
```

```
render = web.template.render('templates/')

urls = (
    '/', 'index',
)

profile = open('profile.yaml')
profile_data = yaml.safe_load(profile)
profile.close()

# Functioning Variables
name = profile_data['name']

tts('Welcome ' + name + ', systems are now ready to run. How can I help you?')

class index:
    def GET(self):
        return render.index()

    def POST(self):
        x = web.input(myfile={})
        filedir = os.getcwd() + '/uploads' # change this to the directory
        you want to store the file in.
        if 'myfile' in x: # to check if the file-object is created
            filepath=x.myfile.filename.replace('\\','/') # replaces the
            windows-style slashes with linux ones.
            filename=filepath.split('/')[-1] # splits the command and
            chooses the last part (the filename with extension)
            fout = open(filedir +'/'+ filename,'w') # creates the directory
            where the uploaded file should be stored
            fout.write(x.myfile.file.read()) # writes the uploaded file to
            the newly created file.
            fout.close() # closes the file, upload complete.
        os.system('python main.py ' + filename)

if __name__ == "__main__":
    app = web.application(urls, globals())
    app.run()
```

You have to import the os, yaml, and web modules. You call the web.template.render() function because you are using web.py's templating engine. This function takes the location of the templates as an argument. Next you specify the list of URLs used in the application. You then define the index class, which contains the back-end code for the index page. The GET() function handles the rendering of the index page.

The POST() function handles the file upload using web.py's form-handling technology and saves the upload to the uploads folder. The comments explain the functionality of each line of code in the function. Finally, the file that is uploaded by the user is passed on to the main.py file as an argument. The Python file is opened from the terminal using the os.system() command.

Now you need to edit main.py to handle the WAV file that is passed as an argument to it. Make the following edits:

```python
import sys

voice_file = os.getcwd() + '/uploads/' + sys.argv[1]

def main(voice_file):
    r = sr.Recognizer()
    with sr.WavFile(voice_file) as source:
        audio = r.record(source)

    try:
        speech_text = r.recognize_google(audio).lower().replace("'", "")
        print("Melissa thinks you said '" + speech_text + "'")
    except sr.UnknownValueError:
        print("Melissa could not understand audio")
    except sr.RequestError as e:
        print("Could not request results from Google Speech Recognition
        service; {0}".format(e))

    play_music.mp3gen(music_path)
    imgur_handler.img_list_gen(images_path)

    brain(name, speech_text, music_path, city_name, city_code, proxy_
    username, proxy_password, consumer_key, consumer_secret, access_token,
    access_token_secret, client_id, client_secret, images_path)

main(voice_file)
```

Let's go through these edits. Using os.getcwd(), the location of the uploads folder, and sys.argv[1] (the name passed via the command line), you can retrieve the WAV file from its location. The main function takes voice_file as an argument. As you may have noticed, you change the code to accept voice input from the WAV file using sr.WavFile() rather than the sr.Microphone() function. You now use the WAV file as the audio source. Do not forget to create the uploads directory in the root by typing the following command in the terminal:

```
$ mkdir uploads
```

Now you need to accept the audio input via the web browser so you can save that file as a WAV file. You use Chris Wilson's Apache Licensed Code to do that. I am not including the JavaScript files in this book for the sake of brevity, but I highly recommend that you go through the code in depth, to get a greater understanding of how the voice is recorded efficiently using a web browser.

Another note: the images used for this project were obtained from iconarchive.com, and their respective authors have approved their commercial use. The JavaScript files are readymade and lengthy. You can obtain all the static files from https://github.com/Melissa-AI/Melissa-Web/tree/master/static.

Next, create the index.html file in the templates folder:

```
<!doctype html>
<html>
<head>
        <meta name="viewport" content="width=device-width,initial-scale=1">
        <title>Melissa - Web Version</title>
    <link rel="stylesheet" href="http://maxcdn.bootstrapcdn.com/
bootstrap/3.3.5/css/bootstrap.min.css">
    <script src="https://ajax.googleapis.com/ajax/libs/jquery/1.11.3/jquery.
min.js"></script>
    <script src="http://maxcdn.bootstrapcdn.com/bootstrap/3.3.5/js/bootstrap.
min.js"></script>
        <script src="../static/audiodisplay.js"></script>
        <script src="../static/recorder.js"></script>
        <script src="../static/main.js"></script>
        <style>
        html { overflow: hidden; }
        body {
                font: 14pt Arial, sans-serif;
            background: url(../static/img/bg-sky.png);
            background-repeat: repeat !important;
            background-attachment: fixed;
                display: flex;
                flex-direction: column;
                height: 100vh;
                width: 100%;
                margin: 0 0;
        }
        canvas {
                display: inline-block;
                background: #202020;
                width: 95%;
                height: 45%;
                box-shadow: 0px 0px 7px blue;
        }
        #controls {
                display: flex;
                flex-direction: row;
                align-items: center;
                justify-content: space-around;
                height: 20%;
                width: 100%;
        }
```

```
#record { height: 13vh;}
#record.recording {
        background: red;
        background: -webkit-radial-gradient(center, ellipse cover,
        #ff0000 0%,lightgrey 75%,lightgrey 100%,#7db9e8 100%);
        background: -moz-radial-gradient(center, ellipse cover,
        #ff0000 0%,lightgrey 75%,lightgrey 100%,#7db9e8 100%);
        background: radial-gradient(center, ellipse cover, #ff0000
        0%,lightgrey 75%,lightgrey 100%,#7db9e8 100%); opacity: 0.5;
}
#save, #save img { height: 10vh; }
#save { opacity: 0.35;}
#save[download] { opacity: 1;}
#viz {
        height: 80%;
        width: 100%;
        display: flex;
        flex-direction: column;
        justify-content: space-around;
        align-items: center;
}
@media (orientation: landscape) {
        body { flex-direction: row;}
        #controls { flex-direction: column; height: 100%; width: 10%;}
        #viz { height: 100%; width: 90%;}
}

        </style>
</head>
<body>
        <div id="viz">
                <canvas id="analyser" width="1024" height="500"></canvas>
                <div id="melissa"><span style="color: #56600FF; text-shadow: -1px
                1px 8px #67C8FF, 1px -1px 8px #67C8FF;">Melissa</span></div>
                <canvas id="wavedisplay" width="1024" height="500"></canvas>
        </div>

        <div id="controls">
                <img id="record" src="static/img/mic.png" onclick="toggleRe
                cording(this);">
                <a id="save" href="#"><img src="static/img/save.png"></a>
                <a data-toggle="modal" data-target="#myModal" href="#"><img
                src="static/img/upload.png"></a>
        </div>
```

```html
<div id="myModal" class="modal fade" role="dialog">
  <div class="modal-dialog">

    <!-- Modal content-->
    <div class="modal-content">
      <div class="modal-header">
        <button type="button" class="close" data-
        dismiss="modal">&times;</button>
        <h4 class="modal-title">Give Melissa The Audio File</h4>
      </div>
      <div class="modal-body">
        <form method="POST" enctype="multipart/form-data"
        action=""><input type="file" name="myfile" /><br/><input
        type="submit" value="Submit"/></form>
      </div>

    </div>

  </div>
  </div>
</body>
</html>
```

First you include all the necessary scripts and CSS files for adding the functionality and styling, respectively. Then comes the canvas element, which displays the waveform simulation of the sounds detected by the microphone. The other canvas element represents the waveforms of the recorded audio from which you want to create the WAV file.

You then have the controls section, which contains a button to toggle starting and stopping recording the user's voice. The next button saves the WAV file that has been recorded to your system. The third button opens a bootstrap modal dialog that gives the user the option to upload the file and submit it to Melissa. As soon as the file is uploaded, the back-end code runs Python main.py audio_file.wav in the terminal using the os.system() function. The WAV audio file is then sent to Google Speech Recognition for conversion of the speech to text. The rest of the machinery works exactly as it did before.

Although I haven't included the static files here, I would still like to discuss the audiodisplay.js file:

```javascript
function drawBuffer( width, height, context, data ) {
    var step = Math.ceil( data.length / width );
    var amp = height / 2;
    context.fillStyle = "silver";
    context.clearRect(0,0,width,height);
    for(var i=0; i < width; i++){
        var min = 1.0;
        var max = -1.0;
        for (j=0; j<step; j++) {
            var datum = data[(i*step)+j];
```

```
        if (datum < min)
            min = datum;
        if (datum > max)
            max = datum;
    }
    context.fillRect(i,(1+min)*amp,1,Math.max(1,(max-min)*amp));
    }
}
```

In the `drawBuffer()` function, if you give the width, height, context, and data as arguments, then, based on the sound levels, it creates a brilliant graph of rectangles side by side. This code, like all the other JavaScript files, was written by Chris Wilson.

This concludes the construction of the front end of Melissa's system—you have a user interface, ready for use. You can start Python's `web.py` server by entering the following command on your terminal:

```
$ python web-gateway.py 127.0.0.1
```

This command notifies you that the server is running on `http://127.0.0.1:8080/`. Go to your web browser (use only Firefox or Chrome) and open the URL, and you will see a web interface like that shown in Figure 8-2.

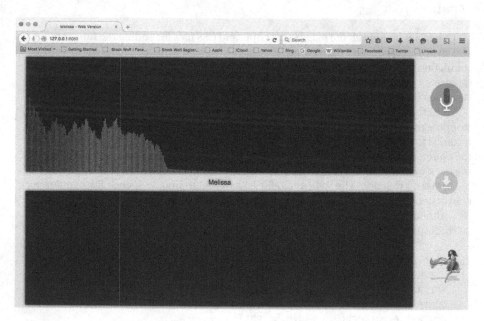

*Figure 8-2.* *Melissa's web interface*

As you can see, the upper panel shows a cool simulation of the sound it receives in real time. (I coughed while taking the screenshot.) Now it's time to record something using your web interface. Say "How are you?" Figure 8-3 shows the waveform of my recorded command in the second canvas element.

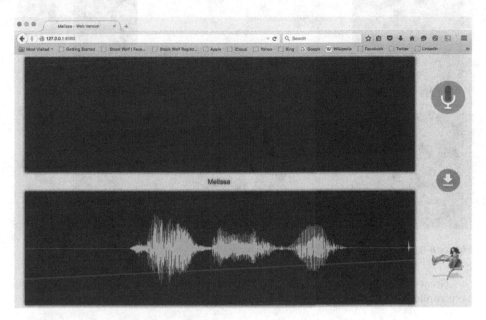

**Figure 8-3.** *Recording audio via the web interface*

I deliberately paused between the words *How*, *are*, and *you*. You can see that there are three main areas of distortion in the waveform indicator (the straight line indicates silence). Save the file using the Save button; it is saved with the name myRecording01.wav. Click the button with Melissa's logo (yes, that's Melissa's logo) to upload the WAV file and submit it. Figure 8-4 shows the modal dialog that gives you the option to upload the file and submit it to the Python back end.

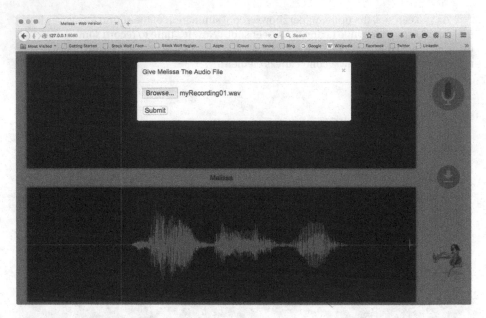

*Figure 8-4.* *Uploading the audio file to the server*

Great—you have successfully created Melissa's easy-to-use web user interface! You have applied existing libraries and your knowledge of Python to develop this system.

# Exercises

This web interface needs some functionality to accept input in the form of text. The option of submitting input from a text field can be incorporated in the modal dialog.

The current workflow for Melissa's front end requires too many clicks. This many clicks for a single command may be frustrating for a user. Your exercise is to devise a new workflow that requires fewer clicks and hence is more user friendly.

# Summary

In this chapter, you developed a web interface for your virtual assistant software, Melissa. You learned to use a third-party JavaScript library for recording a voice using a web browser. You wrote the back-end code using Python's web.py module and learned to work using WAV files. Finally, you wrote the front-end code in HTML and integrated all the pieces of code together to create a working web-based interface for Melissa.

In the next chapter, you get Raspberry Pi running and integrate your software to work in its operating system, Raspbian. You also see how this proof-of-concept software can be scaled to make a full-fledged assistant. The chapter goes through the various enhancements you can make to make Melissa and various sample use cases for the virtual assistant.

# CHAPTER 9

# Integrating the Software with Raspberry Pi, and Next Steps

In this chapter, you learn to get a Raspberry Pi running and integrate your software to work in its operating system, Raspbian. You see how this proof-of-concept software can be scaled to make a full-fledged assistant, and you go through various enhancements and use cases for Melissa.

To this point, you have successfully developed a virtual assistant that listens to you, understands what you say to some extent, and speaks back to you. It can also do a lot of useful things for you, such that tell you the time and weather, tweet, upload pictures, play music, and so on. You have Melissa running successfully on OS X and Linux. Now it's time to set up Melissa in a Raspberry Pi so that she can contribute to making the Internet of Things (IoT) smarter.

First you need to set up your Raspberry Pi (RPi). Even if you don't have a RPi yet, you should still go through this chapter; it will broaden your views on how you can scale Melissa to make her more useful and how you can employ Melissa in different scenarios to make your devices smarter.

## Setting Up a Raspberry Pi

If you have a RPi ready, read about its configuration on the official RPi web site. As you have probably noticed, the RPi comes as a bare-bones board. You need accessories such as a 5 V / 2 amp output micro-USB power adapter, a microSD card for installing the operating system, and an Ethernet cable for connecting the RPi to either your system or your router.

© Tanay Pant 2016
T. Pant, *Building a Virtual Assistant for Raspberry Pi*, DOI 10.1007/978-1-4842-2167-9_9

Once you have these accessories ready, go ahead with the task of installing Raspbian on the microSD card:

- *Linux:* If you own a Linux system, you can install Raspbian on the microSD card using the instructions provided at https://www.raspberrypi.org/documentation/installation/installing-images/linux.md.

- *Windows:* If you own a Windows system, you can install Raspbian on the microSD card using the instructions provided at https://www.raspberrypi.org/documentation/installation/installing-images/windows.md.

- *OS X:* If you own a Mac system, you can install Raspbian on the microSD card using the instructions provided at https://www.raspberrypi.org/documentation/installation/installing-images/mac.md.

This will take a while to complete. Once you have the operating system installed, boot up the RPi (connect the power adapter). If you do not have a spare display, keyboard, and mouse, set up VNC on the RPi via ssh, as described next. You can get the IP address of your Raspberry Pi from your router's administration page under the DHCP Clients section (see Figure 9-1).

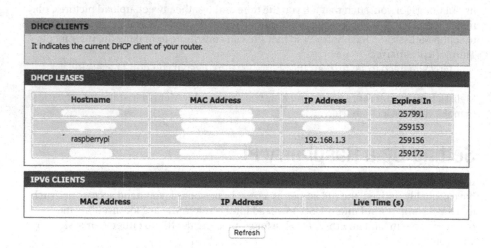

*Figure 9-1.* *Getting the RPi's IP address from the router*

To establish a SSH connection from your system to the RPi, enter the following command on your terminal:

```
$ ssh pi@ipaddress
```

If the connection is successful, it will ask for a password; the default is raspberry.

Now you can install tightvncserver on the system by entering the following command on your terminal:

```
$ sudo apt-get install tightvncserver
$ tightvncserver
```

This asks for an optional view-only password. Start the server by entering the following command after installation:

```
$ vncserver :0 -geometry 1920x1080 -depth 24
```

Next, install software such as a VNC viewer to connect to the VNC server run by the RPi. You can download VNC software for your platform from https://www.realvnc.com/download/. Figure 9-2 shows Real VNC Viewer running on my system and asking for details to connect to the VNC server.

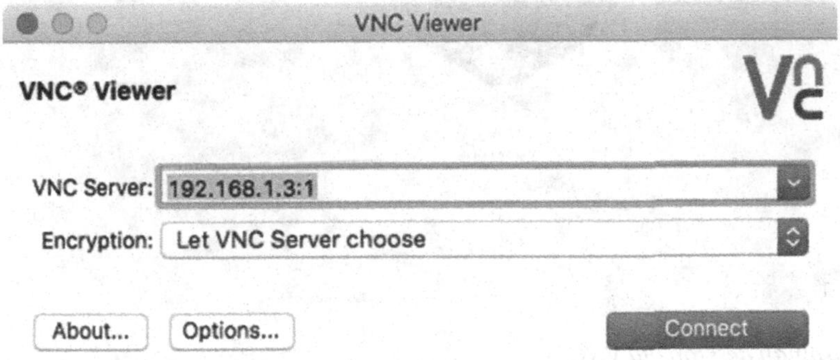

*Figure 9-2.* *VNC Viewer login screen*

This connects you to the RPi VNC server. You see a graphical interface of Raspbian running on the RPi. Figure 9-3 shows Raspbian on my Mac via the VNC connection.

**Figure 9-3.** *Raspbian desktop via VNC*

You can also connect to your VNC server via a VNC viewer on your mobile.

# Setting Up Melissa

To set up Melissa, you first have to repeat all the steps like installing third-party utilities such as PortAudio, PyAudio, espeak, and mpg123. You can find the entire list of things you need to install at https://github.com/melissa-ai/melissa-core.

Now you need to transfer your code repository from your local development environment to RPi. I recommend that you fork the repository linked in the previous paragraph and clone it with the help of git. I entered the following command on my RPi terminal to clone Melissa:

```
$ git clone https://github.com/Melissa-AI/Melissa-Core.git
```

There is another advantage to setting up Melissa by cloning the repository. You can install all the pip modules you have on your local system by entering the following command:

```
$ pip install -r requirements.txt --allow-external pywapi --allow-unverified
pywapi
```

If you are transferring Melissa from your local environment, you must export a list of the Python modules you have installed via pip. You can export all this information to a text file by entering the following commands on your terminal:

```
$ pip freeze > requirements.txt
$ cp profile.yaml.default profile.yaml
$ cp memory.db.default memory.db
```

Once you have successfully set up your development environment, open profile.yaml to customize the file and add details about yourself. Then you can shift your codebase to RPi and install the modules using the method described earlier. You should now be able to run Melissa on Raspbian. If you get any error messages, you may be missing a component. Try to debug Melissa using the error messages provided by Python's interpreter.

# Adding New Components to the Raspberry Pi

You should add some components and accessories to your RPi to work with Melissa. First, you should purchase a case for the RPi. You can find cases on e-commerce web sites like Amazon. I bought a transparent case for my Raspberry Pi, as shown in Figure 9-4.

*Figure 9-4. Raspberry Pi with a transparent case*

It is very important to put your RPi in a case if you power on the RPi while holding it in your hand, because skin conducts electricity. Static electricity can potentially damage semiconductors on the board and kill the RPi.

You also need a microphone so that you can give commands to Melissa, who now resides in the RPi. Microphones that connect via a USB cable are available on Amazon; you can check for compatibility with the RPi in the product's description before purchasing it.

Another thing you need is either earphones or speakers to connect to your RPi so you can hear Melissa's responses to your queries.

# Making Melissa Better Each Day!

This section talks about how you can add new and more complex functionalities to make Melissa better each day. The preferred way to do this is to fork the official repo, open an issue for discussion under new features (this will save you some time if someone else is already working on the issue and will let people decide if they want to help you on a complicated issue), create a new branch for working on your module, and start to work on it. After you have finished building your feature, open a pull request referencing the issue you created earlier. After testing your feature, you can merge it into the official codebase.

The goal is to obtain as many e-karma points as you can by contributing to Melissa's repository. Remember, 1 green box = 1 e-karma point. You can help improve this community-driven, completely open source initiative and make it one of the best virtual assistants in the world. Let's go through some sample features/subprojects that you can own and that would make a big impact on Melissa's functionality.

## Windows Compatibility

Melissa currently supports only OS X and Linux systems. It would be great to include compatibility for Windows as well. That would require going through the entire install methodology in Melissa's repository and documenting changes. It would also require changing the code where the `sys.platform()` function is called and adding references to the Windows platform.

This is one of the easiest yet most important issues that needs to be taken care of in Melissa's repository. It would help you run Melissa on Windows IoT on your Raspberry Pi.

## Tests

No code repository is truly professional unless it has tests built for it. Melissa currently does not have any tests for her health check. These tests would be run by contributors before submitting their pull requests. They might include checking for errors, trailing blank spaces, PEP-8 guidelines conformation, and much more.

You could make a package of tests by creating a `test` directory in the root folder of the project. Or perhaps a test package is not be a good idea at the beginning, but tests should be put in a different directory of their own.

# Vision

Melissa currently accepts voice input from the user, as discussed in the workflow in Chapter 1. It would be great if Melissa could gather information using camera(s). You could use OpenCV for this purpose to add functionalities such as detecting whether a room is empty, counting how many people are in the room, recognizing faces, converting text in photos to strings, and so on.

This would redefine Melissa's current workflow. Figure 9-5 demonstrates how Melissa's workflow might look in the future.

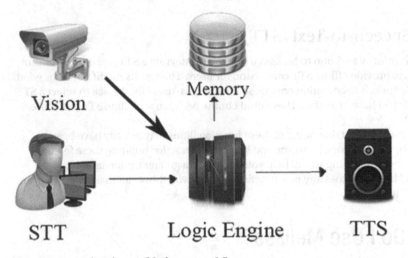

*Figure 9-5. Melissa's possible future workflow*

Melissa would then be able to gather data using vision. This vision feature would go in the SenseCells package, and functionalities built on it would reside in the GreyMatter package.

# Multi-Device Operation

Wouldn't it be cool to have two instances of Melissa running at the same time on different devices but communicating with each other via a server? This would require another piece of software running on a server that handles such requests for devices and connects them using keys that can be requested by a user.

This is easier said than done. It would require quite a lot of programming to build the code for the cloud-based server as well as additions to Melissa-Core for handling the requests made to and from by the server code.

## Native User Interface

Two things that determine the success of software are its usability and looks. Currently Melissa works only via the command line. You built a web interface, but it doesn't have the best workflow, and it is dependent on the user running the Python web-gateway.py file and the voice engine from the command line.

I would like to have a user interface for Melissa that uses the widely used UI Framework (still under discussion) for Python. This would help users interact with Melissa more easily. This is quite a task and would definitely require some time to construct, but it is a very high priority for the project to have a good UI.

## Offline Speech-to-Text (STT)

Another high-priority addition to Melissa would be to integrate a STT like either Julius or CMU Sphinx to provide offline STT conversion for users. The results might not be as good as the Google Speech Recognition engine, so you can give users the choice to select a STT from the STTs you have available. They could choose between an offline STT or a more accurate STT.

By the time you read this, some of these functionalities may already have been constructed by contributors. However, you should still practice building these features on your own, because doing so will help you to achieve a greater understanding of the software. Feel free to discuss any new functionality that you think can make Melissa even better via GitHub issues.

# Where Do I Use Melissa?

You may have the following thought: "Everything is cool, but except for R&D and on a laptop, where do I use Melissa?" Good question! Other than using Melissa on your laptop, there are a couple of sample use cases where I think Melissa can be helpful and make your devices and utilities more accessible and impressive.

## Drones

Many people are building drones using Raspberry Pis and drone kits that are readily available in e-commerce stores. By connecting the motors and functionality to an Arduino board and then to a RPi, you can control the drone's movement, direction of flight, and so on using Melissa, your voice-controlled virtual assistant. You can start the drone simply by giving voice commands and tell it to fly, land, or follow you. The possibilities are limitless with a creative mind.

## Humanoid Robots

Humanoid robots are being developed by big corporations as well as individuals. These autonomous robots use software like Melissa to make them more interactive and, well, human like. If you plan to build a humanoid robot, or any robot for that matter, you can integrate it with Melissa and build appropriate functionalities that extend Melissa to handle your robot.

# House-Automation Systems

You can build your own house-automation system with the help of Melissa. In order to build a small-scale replica as a proof of concept for a house-automation system, you can connect LEDs via a breadboard to the general-purpose input output (GPIO) pins, write a Python script to control the LEDs, and hook it up with Melissa.

Or, if you are like me and prefer not to work with electronics, feel free to purchase a USB-controlled RGB LED stick such as blink(1) (http://blink1.thingm.com). It has built-in integration for USB firmware and can be controlled via a Python script. You just have to plug it into the USB, and then you can control the LEDs via the command line and a Python script. To install the blink(1) command-line tool, enter the following commands on your terminal:

```
$ git clone https://github.com/todbot/blink1.git
$ cd blink1/commandline
$ make
$ sudo make install
```

This installs the blink(1) command-line tool on your system and adds `blink1-tool` to your path. Enter the following command on your terminal to see the various flags you can use to operate the device, as well as some examples:

```
$ blink1-tool
```

Here is the output this command gave me when I entered it in the terminal:

```
Tanays-MacBook-Air:~ tanay$ blink1-tool
Usage:
  blink1-tool <cmd> [options]
where <cmd> is one of:
  --list                      List connected blink(1) devices
  --rgb=<red>,<green>,<blue>  Fade to RGB value
  --rgb=[#]RRGGBB             Fade to RGB value, as hex color code
  --hsb=<hue>,<sat>,<bri>     Fade to HSB value
  --blink <numtimes>         Blink on/off (use --rgb to blink a color)
  --flash <numtimes>         Flash on/off (same as blink)
  --on | --white             Turn blink(1) full-on white
  --off                       Turn blink(1) off
  --red                       Turn blink(1) red
  --green                     Turn blink(1) green
  --blue                      Turn blink(1) blue
  --cyan                      Turn blink(1) cyan (green + blue)
  --magenta                   Turn blink(1) magenta (red + blue)
  --yellow                    Turn blink(1) yellow (red + green)
  --rgbread                   Read last RGB color sent (post gamma-correction)
  --setpattline <pos>        Write pattern RGB val at pos (--rgb/hsb to set)
```

```
  --getpattline <pos>        Read pattern RGB value at pos
  --savepattern              Save color pattern to flash (mk2)
  --play <1/0,pos>           Start playing color sequence (at pos)
  --play <1/0,start,end,cnt> Playing color sequence sub-loop (mk2)
  --playpattern <patternstr> Play Blink1Control pattern string
  --servertickle <1/0>[,1/0] Turn on/off servertickle (w/on/off, uses -t msec)
  --chase, --chase=<num,start,stop> Multi-LED chase effect. <num>=0 runs forever.
  --random, --random=<num>   Flash a number of random colors, num=1 if omitted
  --glimmer, --glimmer=<num> Glimmer a color with --rgb (num times)
 Nerd functions: (not used normally)
  --eeread <addr>            Read an EEPROM byte from blink(1)
  --eewrite <addr>,<val>     Write an EEPROM byte to blink(1)
  --fwversion                Display blink(1) firmware version
  --version                  Display blink1-tool version info
and [options] are:
  -d dNums --id all|deviceIds Use these blink(1) ids (from --list)
  -g -nogamma                Disable autogamma correction
  -m ms,   --millis=millis   Set millisecs for color fading (default 300)
  -q, --quiet                Mutes all stdout output (supercedes --verbose)
  -t ms,   --delay=millis    Set millisecs between events (default 500)
  -l <led>, --led=<led>      Set which LED in a mk2 to use, 0=all,1=top,2=bottom
  -l 1,3,5,7                 Can also specify list of LEDs to light
  -v, --verbose              verbose debugging msgs


Examples
  blink1-tool -m 100 --rgb=255,0,255    # fade to #FF00FF in 0.1 seconds
  blink1-tool -t 2000 --random=100      # every 2 seconds new random color
  blink1-tool --ledn 2 --random=100     # random colors on both LEDs
  blink1-tool --rgb 0xff,0x00,0x00 --blink 3  # blink red 3 times
  blink1-tool --rgb '#FF9900'           # make blink1 pumpkin orange
  blink1-tool --rgb FF9900 --ledn 2     # make blink1 pumpkin orange on
lower LED
  blink1-tool --playpattern '10,#ff00ff,0.1,0,#00ff00,0.1,0'
  blink1-tool --chase=5,3,18            # chase 5 times, on leds 3-18

Notes
 - To blink a color with specific timing, specify 'blink' command last:
   blink1-tool -t 200 -m 100 --rgb ff00ff --blink 5
 - If using several blink(1)s, use '-d all' or '-d 0,2' to select 1st,3rd:
   blink1-tool -d all -t 50 -m 50 -rgb 00ff00 --blink 10


Tanays-MacBook-Air:~ tanay$
```

I also created a new module named lighting.py in the GreyMatter folder that contains the following code:

```python
import osfrom SenseCells.tts import tts
def very_dark():
    os.system('blink1-tool --white')
    tts('Better now?')

def feeling_angry():
    os.system('blink1-tool --cyan')
    tts('Calm down dear!')

def feeling_creative():
    os.system('blink1-tool --magenta')
    tts('So good to hear that!')

def feeling_lazy():
    os.system('blink1-tool --yellow')
    tts('Rise and shine dear!')

def turn_off():
    os.system('blink1-tool --off')
```

Now make the following edits/additions to the brain.py file:

```python
from GreyMatter import notes, define_subject, tell_time, general_
conversations, play_music, weather, connect_proxy, open_firefox, sleep,
business_news_reader, twitter_interaction, imgur_handler, lighting

    elif check_message(['feeling', 'angry']):
        lighting.feeling_angry()

    elif check_message(['feeling', 'creative']):
        lighting.feeling_creative()

    elif check_message(['feeling', 'lazy']):
        lighting.feeling_lazy()

    elif check_message(['dark']):
        lighting.very_dark()

    elif check_message(['lights', 'off']):
        lighting.turn_off()
```

This acts as a small-scale replica of how you can control your house lighting, doors, and so on using programming. All you have to do is add this code to a separate file and make the appropriate changes to brain.py. This way, a command like "Smoothen lighting!" could make the your lights turn blue.

You can make many variations of this functionality with just a little programming. For example, you could use this along with the "Party mix!" command you integrated into your software earlier in this book, to set the lights to blink or change color randomly!

## Burglar-Detection System

Using vision integration with OpenCV, you can detect whether someone has entered your house in your absence. Extending that functionality, you can program Melissa to take a picture of the person; alert you by ending a message that someone is in your house, along with their photo; call 911; and sound an alarm.

Many other features can be integrated into such a system. Try brainstorming about it.

# Summary

In this chapter, you learned how to set up a Raspberry Pi and integrate Melissa into it. Then you saw how to continue your learning after you finish reading this book and where you can implement this virtual assistant to make the most of your devices.

In this book, you learned about virtual assistants, famous virtual assistants available on the market, developing a new virtual assistant, and making the virtual assistant speak, listen, and understand what the user says. You then built several modules that let you talk with Melissa and ask her for information such the weather report, definitions from Wikipedia, and the time. You also developed modules with which Melissa can tweet for you, play music for you, save notes for you, and upload images for you.

I strongly encourage you to keep working on Melissa after you finish reading this book. Doing so will reinforce the concepts in your brain. Follow the principle of "Making Melissa Better Each Day!" Together, we can make Melissa one of the best open source virtual assistants in the world.

Stay hungry; stay foolish!

# Index

© Tanay Pant 2016
T. Pant, *Building a Virtual Assistant for Raspberry Pi*, DOI 10.1007/978-1-4842-2167-9

## ■ W, X, Y, Z

# Get the eBook for only $5!

Why limit yourself?

Now you can take the weightless companion with you wherever you go and access your content on your PC, phone, tablet, or reader.

Since you've purchased this print book, we're happy to offer you the eBook in all 3 formats for just $5.

Convenient and fully searchable, the PDF version enables you to easily find and copy code—or perform examples by quickly toggling between instructions and applications. The MOBI format is ideal for your Kindle, while the ePUB can be utilized on a variety of mobile devices.

To learn more, go to www.apress.com/companion or contact support@apress.com.

Printed in the United States
By Bookmasters